Rashda: The Birth and Growth of an Egyptian Oasis Village

Islamic Area Studies

Series Editor

Toru Miura (*Ochanomizu University*)

VOLUME 4

The titles published in this series are listed at *brill.com/isas*

Rashda: The Birth and Growth of an Egyptian Oasis Village

By

Hiroshi Kato
Erina Iwasaki

BRILL

LEIDEN | BOSTON

Cover illustration: View from the top of "old" Rashda. The oldest mosque in Rashda, Atiq Mosque, and the new Rashda are on the left-hand side, with the palm tree orchard of the Bir Sheykh Irrigation District on the right-hand side. Photograph by the authors.

Want or need Open Access? Brill Open offers you the choice to make your research freely accessible online in exchange for a publication charge. Review your various options on brill.com/brill-open.

Typeface for the Latin, Greek, and Cyrillic scripts: "Brill". See and download: brill.com/brill-typeface.

ISSN 2214-6555
ISBN 978-90-04-31617-1 (hardback)
ISBN 978-90-04-31739-0 (e-book)

This book is printed on acid-free paper and produced in a sustainable manner.

Contents

PART 2
Rashda in History

PART 3
Rashda Today

Preface

This book is based on an international and interdisciplinary joint project that has been running since 2004. The authors, Hiroshi Kato (socio-economic history) and Erina Iwasaki (economic sociology), visited Rashda village for the first time in 2005, and selected it as one of the villages to be surveyed in the framework of a joint research project on Egypt's socio-economy between Hitotsubashi University and CAPMAS in Egypt. Since 2009, Dr. Nobuhiro Matsuoka of Chiba University (agricultural meteorology) and Dr. Reiji Kimura of Tottori University (meteorology), joined the project, followed in 2010 by Dr. El-Sayed Zaghloul (hydrogeology and geoarchaeology) and Dr. Salwa Elbeih (remote sensing and GIS applications in civil engineering) of NARSS (National Authority for Remote Sensing & Space Sciences). Some results of our project have been published (Kato and Iwasaki 2008, Kato et al. 2010, 2012, 2014). This book was planned to revise and enlarge the published results of our research, by adding to them the results of new fieldwork and analyses.

Most of the analyses and descriptions are based on data and information collected before the January 25 Revolution of 2011. The revolution had a large impact on the society of small villages located far from Cairo, such Rashda in the peripheral desert region, and has possibly provided the impetus for a new future for the village. Since the revolution, Rashda has seen considerable development, with old houses in the old area of the village being demolished and many new houses and shops being built. Public utilities related to communal life are being rebuilt or renewed. Despite these changes in the landscape of the village, the social structure of village life was not able to change easily after the revolution.

In addition of our research colleagues mentioned above, we are deeply grateful to CAPMAS staff, especially Mr. Ali Ahmad Ali, for arranging the household surveys in Rashda and to Dr. Adel Sharaby of NARSS and Dr. Abd al-Hay Aly Farrag of the Faculty of Science, Asyut University, New Valley, for conducting the fieldwork together in Rashda. We are also indebted to the village chief (*ra'is al-qarya*), current chief Badr al-Din Ali, previous chiefs Fathi Ahmad and Abd al-Maniam Nasr, late Omda Kamal Imam.. We thank all the villagers of Rashda in particular *Hajj* Hasan Umran who provided us with his thoughtful knowledge about the history of Rashda, Mr. Saad Hasan and Mr. Mahmud Sayyid Abid who is in charge of water distribution in the Well No. 3 Irrigation District, and taught us about the history and complicated system of water distribution and irrigation.

Special thanks go to Mr. Gibril Abu al-Saoud. He has been an excellent informant for our research, and perhaps the most crucial person in the development and publication of this book. Without his help, the case study of Rashda would not have been possible.

Our research was mainly financed by a Grant-in-Aid for Scientific Research, Japan Society for Scientific Research (Ministry of Education, Culture, Sports, Science and Technology, Japan), which has supported a number of research projects: "The collection of basic Egyptian socio-economic data and an interdisciplinary analysis based on them" (2004–2007), "Study on the dynamics of Arab social transformation—focus on the collection and analysis of panel data" (2010–2014), "Study of living security mechanism in local society – empirical study of mutual customs in Egypt" (2009–2012), and "Coexistence in a Water-Scarce Society—Irrigation Systems" (2013-present). We deeply appreciate the financial assistance of the Japanese government over the long term for our fundamental academic research.

Transliteration of Arabic

The transliteration of Arabic presents great difficulty in a work such as this book on the study of a specific area in which a great many proper nouns referring to places and people appear. We have adopted the same system of transliteration of Arabic and Arabized words used in the *International Journal of Middle East Studies* (*IJMES*), but without all the diacritic marks, retaining only the ayn and hamza. Regarding the names of places and people, all diacritics, including the marks for the ayn and the hamza, have been omitted. If a place name has a familiar English form, such as Port Said, this is used rather than the Arabic (bur saʿid). Otherwise, the place names are transliterated according to colloquial pronunciation, or written as commonly transliterated by Egyptians themselves. The classical Arabic vocalization for the technical and professional words can be found in the Explanatory Glossary. Finally, for Arabic measures of weights, area and currency, the italic singular is used, although it signifies the plural. Therefore, *feddan* is used instead of *feddans*.

List of Illustrations

All photographs by the author unless indicated otherwise

Maps

Figures

Tables

Graphs

Photos

Explanatory Glossary

adhan Call to prayer. In the private mosques (*masjid ahli*), this is performed by villagers living in the same area by rotation in Rashda today. ⇒ *masjid ahli*

ʿagwa Date paste. This is used for making sweets.

ʿaʾila Aila. Family. This is a social unit based on a household (*usra*), while a household is a residential unit. In Egypt, the family is generally called '*aʾila*' in Arabic. '*Aʾila*' is sometimes translated as 'extended family'. The term '*aʾila* is one of the most important, and frequently mentioned, technical terms in this book. Therefore, we write this as "aila" instead of *ʿaʾila* in the text for convenience. ⇒ *usra*

ʿain Spring. Today, the villagers of Rashda recognize underground water sources according to depth. They are called "wells" (*bir*) if they must be tapped using a drill, or "springs" (*ʿain*) if the water bubbles up naturally and only needs to be cleared from time to time. However, with the disappearance of artesian wells, *ʿain* is now becoming a historical term. We translate the word *ʿain* as "spring" for convenience, because it is frequently mentioned in the text. ⇒ *bir*

ʿain sathi A type of well called a "surface spring". These are popularly recognized as springs that bubble up by themselves. Surface springs, in addition to investment wells, began operating in the 1990s with economic liberalization. They are drilled with permission from the Land Reclamation Fund, and are individually or collectively owned and managed by the cultivator(s). ⇒ *bir istithmari*

ʿain rumani Old well. This literally means "Roman spring". Roman springs are popularly recognized as dating back to the Roman era, although most of them are assumed to be 200 years old or less. They have all dried up now.

amila (*plu. amaʾil*) Share of the right to use water. This is the most important concept in the division of water. It is the unit of time, which equals one day or night (or from sunrise to sunset, or from sunset to sunrise); each *amila* is divided into *qadam*, which usually number 30. The smallest unit is the *habba*, which equals one-quarter of a *qadam*. Thus, one *amila* equals 30 *qadam* or 120 *habba*. This is so even for the district attached to a government well.

ʿarab Bedouins. This literally means the Arabs, but here signifies, as well as *ʿurban*, the Bedouins. ⇒ *ʿurban*

ardabb Dry measure used for grain. Historically, there were various local *ardabb* in Egypt (cf. Cuno 1992:209).

ʿasabiya Brotherhood based on the fraternal relationship.

awlad al-ʿamm/ʿamma Cousins on the father's side (*ʿamm*) and the mother's side (*ʿamma*).

bait Household. This word literally means "house", and in addition to *manzil* it signified so-called households in documents such as population censuses in earlier times. ⇒ *usra, manzil*

bakiya (or *nahr*) A method of *ghamr* (flood) irrigation. Parcels of cultivated land are divided into divisions (*bakiya*) and irrigated by small canals (*qanat taqdim*) inside the parcels. ⇒ *nahr*

baqqala Small grocery shop.

bir (plu. *abyar*) Well. We use the English word "well" instead of *bir* because it is frequently mentioned in the text. ⇒ *ʿain*

bir ahli A type of well called a "local well". These are exploited by local cultivators, who dig them in the traditional way and own and manage them collectively. Most of these wells were initially dug as artesian wells.

bir hukumi A type of well called a "government well". These are the wells that have been opened since the 1960s under the Land Reclamation Scheme of the Gamal Abdul-Nasser era. They were drilled and dug by the General Authority of Desert Development and are currently owned and managed by the Irrigation Authority under the Ministry of Water Resources and Irrigation.

bir al-istiʿawad Substitution well. When a local well dries up and a new well is dug or a new pump is installed the newly dug well is called a "substitution well".

bir istithmari A type of well called an "investment well". Investment wells, in addition to surface springs, are types of wells that began operating in the 1990s with economic liberalization. Investment wells are exploited by one or several individuals ("investors") under a General Authority investment scheme that began in 1995. ⇒ *ʿain sathi*

bir al-sabil Well for drinking water.

birka Drainage pond. *Birka* literally means "small lake", and in an Egyptian village it meant the small pond communally owned by villagers in old times.

birsim baladi Egyptian clover. Fodder crop. There are four varieties of Egyptian clover in Egypt: *misqawi, saʿidi, fahl* and *khadrawi*. In Rashda, the most common variety is *misqawi*. This is common in Lower Egypt, where there is abundant water in winter. It grows rapidly, and can be cut four to eight times in a season. ⇒ *birsim hijazi*

birsim hijazi Alfalfa. This is a perennial fodder crop. Alfalfa, or lucerne, is a common crop used to provide fodder for cattle. It is said that if *birsim baladi* is the "king" of fodder crops, *birsim hijazi* is the "queen". It has less nitrogen than *birsim baladi*, but is much more suitable for dried hay. Probably for this reason, *birsim hijazi* is exported to Gulf countries in bales. ⇒ *birsim baladi*

bur Fallow land. The term literally means "uncultivated land", and is used to signify fallow land in the rotation of cultivation.

dabit Military officer.

dar al-munasabat Gathering place. This is used for occasions such as weddings and funerals. In Upper Egypt, the word *duwwar* is popularly used to designate a gathering place.

darb Blind alley or quarter of a village. This is also called a *hara, sikka* or *zuqaq*.

dawra Rotation. In particular, this signifies the rotation of water distribution and crop cultivation.

effendi Turkish title (*affandi* in Arabic) for a non-native notable.

faliq asfar Type of date palm.

feddan or *faddan* Areal measure. Nowadays it is equal to 1.038 acres.

fellahin or *fallahin* Peasants. This is the term used instead of the Bedouins (*ʿarab, ʿurban*). ⇒ *ʿarab, ʿurban*

ful Fava bean.

gafir Village watchman. He is responsible for security issues in the village under the *umda* (village chief). He is now a public servant.

gawafa Guava. A fruit.

ghamr Flood irrigation. The water floods from the main or branch canal into the parcel until the entire field is covered. This is called *ghamr sathi*, or *bakiya*, or *ghamr kulli* or *batti*, according to the method of irrigation inside the basin. ⇒ *bakiya*

ghayt and *ghuta* Field of cultivated land.

ghawwasin Divers who restored springs in earlier times.

habba Unit of measurement for *amila* in this case. 1 *amila* = 30 *qadam* = 120 habba. ⇒ *amila*

habbaka Roof of an old spring (*ʿain*).

hajj Honorific title of one who has performed the pilgrimage (*hajj*) to Mecca.

hakim Legal administrator or governor of the *hukm* system. In those days, he probably played a role equivalent to that of the *qadi* (judge). ⇒ *hukm, qadi*.

halaf Long and strong grasses (*hasha'ish*) used for fencing to keep out sand.

haqq Right (*haqq*) to use the water of a spring or well in the water distribution.

haml Weight measure for rice. 1 *haml* = 4 *wiba* = 40 *misha*.

hara Blind alley. ⇒ *darb*

hawd Basin. This is the unit of cultivated land for irrigation and taxation purposes.

hissa A quarter of a village.

hukm Administrative system in the Ottoman age. *Hukm* meant the administrative hierarchy, at the top of which was the *hakim* (governor). ⇒ *hakim*

ʿid al-Adha Feast of Immolation.

ijar Lease.

imam Prayer leader of a mosque. Traditionally he was elected from village notables, but is now appointed by the Ministry of the Waqf.

infitah Open-door policy in the 1970s.

ishtirak Partnership.

ism Given or first name.

'izba Small settlement. This is a settlement attached to a village. *'Izba* historically signifies a small settlement of peasants who provided labour for farms in Lower Egypt.

ja'ala Jaala. Custom or contract of co-operation. This is, in a sense, a type of *musharaka*. In fact, people involved with a jaala are called *shuraka'* (sing. *sharik*), as well as *muja'alin*. The terms *Shuraka' and muja'alin* mean "partners". The term *ja'ala* is one of the most important and frequently mentioned technical terms in this book. We write it as *jaala* instead of *ja'ala* in the text for convenience. ⇒ *musharaka, sharik, muja'alin*

ja'ala al-nizara Type of jaala for an inspector. The inspector receives an *amila* (the right to use water) in compensation for inspecting water distribution.

jam'iya Association. In Egypt, every kind of organization is called a *jam'iya* and is obliged to register with the Ministry of Foreign Affairs.

jisr Dike or bank of canals.

jurn Threshing place. In the Oasis region, this is usually an open space that is not owned by anyone, in the vicinity of a village. Villagers own parts of it by *wad' al-yad* ("putting the hand"). ⇒ *wad' al-yad*

kashf Register of water distribution. This indicates the share of each owner in the local well (*bir ahli*). It is kept by the Chief of Water (*rais al-ma'*) who manages the water schedule and mediates any problems with irrigation. ⇒ *bir ahli, rais al-ma'*

kashf al-mukalafa Register of agricultural lands and real estate maintained by the Real Estate Tax Department.

kharij al-zimam Land outside the *zimam*, which was unsurveyed and unregistered in the Survey Department Books and Registers. The term literally means "outside of the *zimam*" and the land is now owned by the government. ⇒ *zimam*

khatt Parcel of cultivated land.

kubar (*sing. kabir*) Elders. Elders are sometimes called *shaykh* (notables). ⇒ *shaykh*

laqab Last name or surname. In our research, this is used as a term to signify a family name. (*laqab al-'a'ila*). ⇒ *laqab al-'a'ila*

laqab al-'a'ila Family name in our household survey.

ma'dhun or (*al-*)*ma'dhun* (*al-*)*shar'i* Religious official. In Islamic law, the title of *ma'dhun* refers to the official authorized by the *al-qadi* to perform civil marriages. However, in the local context of the Dakhla Oasis, he also had a role as the secretary of the religious court (*qalamiya al-mahkama*) to draw up official documents not only for civil marriages but also for civilian affairs such as water distribution. The *ma'dhun* was not appointed by the authorities but was selected according to local customs (*'urf*). The post of *ma'dhun* has usually been assumed by members of prominent families. ⇒ *qalamiya al-mahkama, 'urf.*

madina City. The urban area, as opposed to the *rif* (rural area). ⇒ *rif*

mahkama or (*al-*)*mahkama* (*al-*)*shar'iya* Religious Court.

majlis Council.

majlis al-nizara li-mazraʿa Water management board of an irrigation district.

majlis al-shaʿb al-mahalli Local People's Council. This is a council in an *al-wahda al-mahalliya* (Local Unit), and is elected by the villagers. ⇒ *(al-)wahda (al-)mahalliya*

majlis ʿurfi Customary council. In the traditional system of village administration, the council supported the *umda* (village chief), but officially ended with the establishment of *al-wahda al-mahalliya* (Local Units) in 1970. ⇒ *(al-)wahda (al-)mahalliya*

maʾmur Administrator of *maʾmuriya.* ⇒ *maʾmuriya*

maʾmuriya District of administration.

manzil Household. This literally means "house", and in addition to *bait* the term was used to signify a so-called household in documents such as population censuses in old times. ⇒ *usra, bait*

maqar al-ʿumdiya *Umda* (village head) office. This was usually the house of the *umda.*

maqsam Distributor of water. This controls the water flow at a canal junction.

markaz District of administration, or the central city of the region, or the official authority. It literally means "the centre".

marwa Small canal by which water flows into parcels (*khatt*) of cultivated land after the primary and the secondary canals. ⇒ *khatt*

masjid ahli Private mosque. Such a mosque is established by co-operation between villagers. Unlike the government mosque, financed and constructed by the government, the private mosque is financed and built by the villagers themselves. If villagers want to construct a mosque, they submit an application to the Ministry of Waqf and Security.

mastaba Ridges that form a parcel for irrigation. ⇒ *nahr*

mawat Virgin land not possessed by anyone. It literally means "the death land".

mawquf Waqf Materials. ⇒ *waqf*

midan Square.

misha Weight measure for rice. ⇒ *haml*

mishmash Apricots.

muʿawin Police officer.

mudir Chief of *mudiriya* (governorate).

mudiriya Governorate or province. Administrative division used before a change to *muhafaza* (governorate). The chief of the division was a *mudir.*

mufti Eminent scholar of Islamic law (*ʿulamaʾ*), who is qualified to issue opinions regarding religious affairs.

mullak (sing. *malik*) Owners.

musharaka Co-operation of a reciprocal nature or a kind of partnership. The term literally means "co-operation" or "partnership". *Musharaka* takes diverse forms according to the circumstances, as is understood from the diverse usage of the word *musharaka* in Rashda. *Musharaka* includes three kinds of co-operation. The first type is co-operation among family members. The second type is co-operation based on contracts for share-cropping and leasing (*ijar*). The third type is co-operation in exchange for labour between the villagers. ⇒ *ja'ala, ijar*

musharaka mujamula A type of *musharaka*. The term means "friendly co-operation" without recompense.

musharik (*plu. musharikin*) Partners of *musharaka*. ⇒ *sharik*

mutaja'ilin or *mutaga'ilin* People covered by a *ja'ala* contract. It literally means "the people engaged with the *ja'ala* contract". ⇒ *ja'ala*

muzara'a Share-cropping contract between a landowner and a tenant farmer.

muzari'in Share-croppers.

nahiya (plu. *nawahi*) Village. In old documents such as the population censuses of 1882 and 1897, a village was called a *nahiya*.

nahr Furrow irrigation (or *bakiya*). It is called "furrow irrigation" because a parcel is formed by ridges (*mastaba*), and there are water passages or furrows (*nahr* or *khulayja*) between these ridges. ⇒ *mastaba, bakiya*

najjarin or *naggarin* Local professional diggers of wells. The term literally means "carpenters", who dug the traditional artesian wells in the old days. ⇒ *bir ahli*

nazla Small settlement. It also means a quarter of "new" Rashda.

nisf fidda Unit of currency. The *nisf fidda* or the "silver para" literally means "half silver", and was the most commonly used coin in the eighteenth century (cf. Cuno 1992:211).

qabila Tribe. Larger kinship group than *'a'ila*. ⇒ *'a'ila*

qada' Judicial system in the Ottoman age. The *qada'* signified the judicial hierarchy, at the top of which was the *al-qadi* (judge) of the *mahkama* (religious court). ⇒ *al-qadi, mahkama*

qadam Unit of measurement for *amila* in this case. One *amila* = 30 *qadam* = 120 habba. ⇒ *amila*

qadi Judge of the religious court (*mahkama*).

qalamiya al-mahkama Secretary of the religious court. He drew up official documents not only for civil marriages, but also for civilian affairs such as water distribution. ⇒ *mahkama*

qanat taqdim Small canals by which the parcels of fields are irrigated.

qarya Village. Smallest administrative unit in rural areas.

qirat Karat. In Rashda, this has two meanings. The first is the unit that signifies the share of a total of 24 *qirat*. The second is the share of water distribution. One *qirat* is considered to be 100 cubic metres.

qirsh Piastre. This was the currency of Egypt until 1834, and was subdivided into 40 *para*, each of 3 *akçe*.

qism Subdivision of a city (*madina*), that is, a district. ⇒ *madina, shiyakha*

ra'is al-ma' Responsible for the water allocations and distribution from the local wells. *Ra'is al-ma'* means "the chief of water", nominated by the elders and those with experience. He keeps a register book (*kashf*), manages the water schedule, and mediates any problems with irrigation. ⇒ *kashf*

rif Country or rural area, as opposed to the city or urban area (*madina*). ⇒ *madina*

sabkha Salt flat.

sahm Areal unit of measurement. This is 7.293 m^2 in Egypt. However, in Rashda, it signifies the a certain amount of "share" of irrigation hours of water.

sa'idi A variety of date palm. There are many varieties of date palm, but in Rashda, and in the Western Desert in general, the most common varieties are *sa'idi* and *tamar*. *Sa'idi* is a popular variety in Egypt for the soft and sweet taste of the dates, which are sold in markets. They are also used to make date paste (*'agwa*). *Hayani* is also a common variety in Egypt, but these dates are rarely grown in Rashda. They are red-black in colour and are mostly for domestic sales rather than export. ⇒ *'agwa*

saqiya A device for lifting water for irrigation. It uses buckets, jars, or scoops fastened to a vertical wheel.

sant A type of tree, *Acacia nilotica*. It grows abundantly around the oasis, and its wood is extraordinarily durable in water.

sayyid Honorific title for gentlemen.

shari' Open main street as opposed to *darb* (blind alley). ⇒ *darb*

shari'a Islamic law.

sharik (*plu. shuraka'*) Partners of *sharika*. ⇒ *musharikin*

sharika ⇒ *musharaka*

shaykh or *sheykh* (*plu. mashaykh*) Notable. *Shaykh* is the honorific title for village notables and elders. ⇒ *kubar*

shaykh al-balad Village shaykh. The *Shaykh al-balad* was the traditional leader of village society and represented the villagers to the state. He was the representative of the *hissa* (village quarter) and of the largest family (*'a'ila*) in the *hissa*. Therefore, the *hissa* was named for the chief of its largest family, and the village administration was managed by consensus of several *shaykh*. ⇒ *hissa*

shiyakha Smallest subdivision of a city (*madina*), which is a town. ⇒ *madina, qism*

shuna Storage building. Used to store tools and goods for agriculture.

sikka Blind alley. ⇒ *darb*

sufi Member of *tariqa* (Sufi order).

tahuna Mill.

tamar al-qa'aqa' A variety of date palm. *Tamar* are usually eaten after being sun dried, and are mostly for home consumption.

tariqa Sufi order.

tasharrud Vagabond, one of the marginal people, as opposed to an Egyptian national.

tut Mulberry.

ʿumda Umda. Village chief. Before the introduction of this post in 1880/81, the village official was called the *shaykh al-balad* (village shaykh). He was the representative of the *hissa* (village quarter) as well as of the largest family (*ʿaʾila*) who lived in the *hissa* concerned. Therefore, several *shaykh* were appointed in one village to administer the village. In contrast, one umda was appointed in every village. While the *shaykh al-balad* was the traditional leader of village society and represented the villagers to the state, the umda was literally responsible for village control under the direct supervision of the state, although both of them were village officials standing between the state and village society. The term *ʿumda* is one of the most important and frequently mentioned technical terms in this book. We write it as "umda" instead of *ʿumda* or "village chief" in the text for convenience. ⇒ *shaykh al-balad, hissa, shaykh al-ʿaʾila*

ʿumla sagh miri This literally means "the governmental pure money", which implies that the coin concerned is the legal Egyptian currency.

ʿurban Bedouins. This term is used to distinguish them from peasants (*fellahin*).

ʿurf Customary law or local customs.

usra (*plu. usar*) Household. This is, by definition, a residential unit where individuals share consumption, and is usually called an *usra* in Arabic, although they were called *bait, manzil* and other terms in documents such as population censuses in old times. ⇒ *bait, manzil*

wadʿ al-yad Possession. This term literally means "putting the hand". The act of farmers "putting their hands" on the earth is a culturally, but not legally, accepted method of acquiring land.

waha Oasis.

wahda mahalliya Local Unit. A local unit is the administrative unit composed of a "mother village" and some other villages. Its head is the Chief of the Local Unit (raʾis *al-wahda al-mahalliya*).

waqf Religious endowment.

waqfiya *Waqf* document.

waqif Donor to the *waqf.*

warsha Small workshop.

wiba Weight measure for rice. ⇒ *haml*

zakat al-fitr Charity given to the poor at the end of the month of Ramadan, generally in the form of food. Its main purpose is to provide the poor with a means to celebrate the festival of breaking the fast (*ʿid al-fitr*) along with other Muslims.

zakat al-mal This means obligatory "alms" or "alms upon wealth". Every Muslim possessing the designated minimal amount of wealth, *nisab*, for the full cycle of a lunar year, is required to contribute a fixed portion of their wealth to help the needy.

zariba Animal shed. These are seen in many places surrounding the residential area of Rashda.

zimam Land surveyed and recorded in the Survey Department Books and Registers. The term is used in Rashda to signify the subdistricts of the irrigated land. ⇒ *kharij al-zimam*

zuqaq Blind alley. ⇒ *darb*

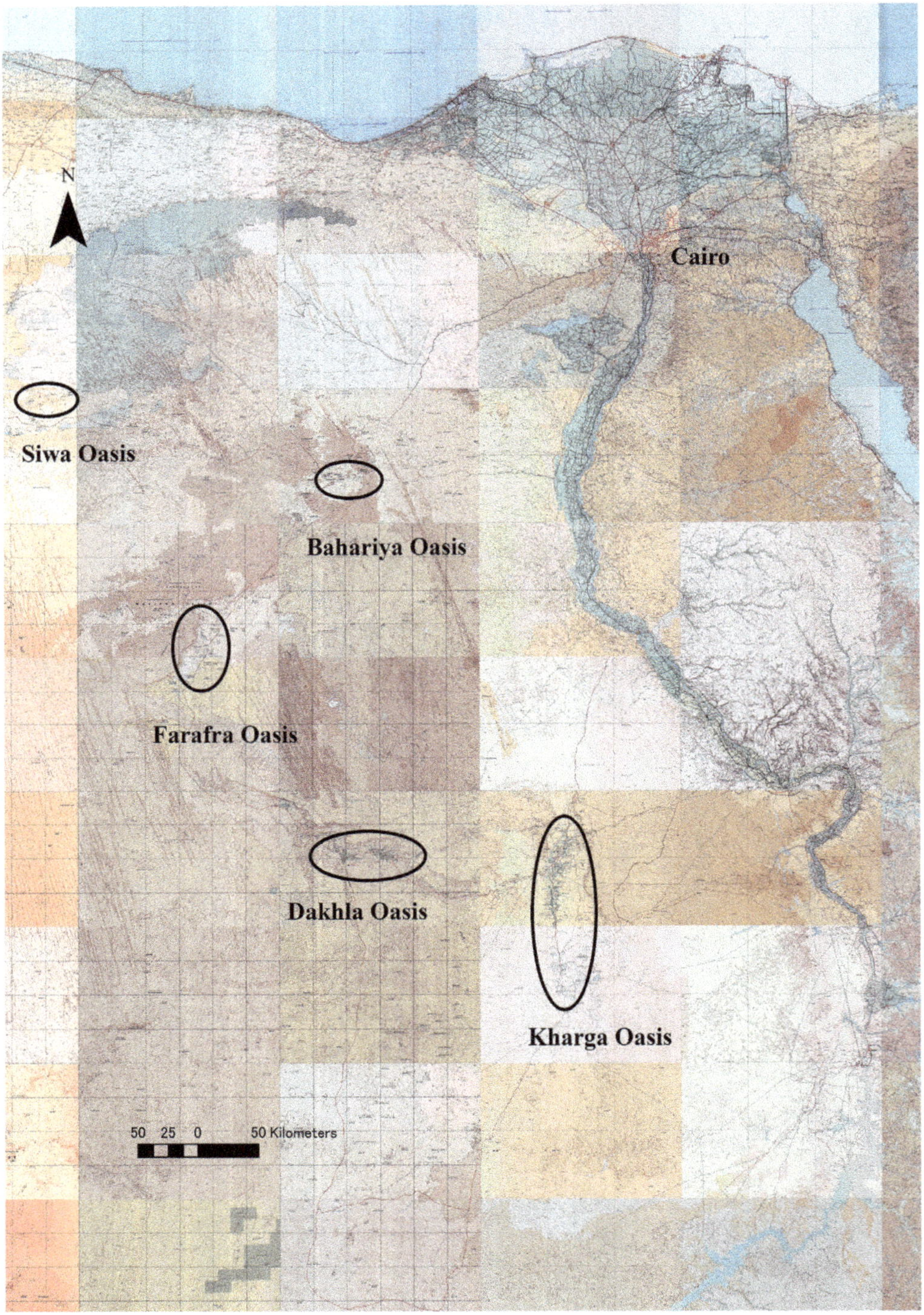

MAP 1 *Dakhla and Kharga Oases in the Western Desert. Source:* 1:250,000 scale maps (1986), Egyptian General Survey Authority.

Introduction

Scope of the Research

The Desert as a Negative Image of Egyptian Society

Egypt is composed of four administrative areas: the Urban Governorates, Lower Egypt, Upper Egypt, and the Frontier Governorates. The Nile is divided into two branches, which form a delta at Cairo. Lower Egypt is the northern part, from Cairo to the Mediterranean. Upper Egypt is the southern part, from Cairo to the border between Egypt and Sudan. In 2014, the Frontier Governorates included two governorates in Sinai (North and South Sinai) and two governorates in the Western Desert (Matruh and New Valley).[1]

Egypt is well known for being a society dependent on the Nile for water. As such, Egypt has emerged and been described as a centralized society. This view has influenced the literature on Egyptian society. A common view of Egyptian history is that it is a centralized state in which the capital city of Cairo plays a dominant role in every aspect of its inhabitants' lives.

This image has also been characterized by a prevailing dualism. This view is concerned with centre–local relations; that is, the contrasts between metropolitan Cairo and other local communities. One outcome of this view is the neglect of the regional perspective. This neglect of regional differences in the spatial configuration is deeply connected with the historical view of the progressive and unitary development from the traditional to the modern in Egyptian society.

In the conventional view of Egypt, its centre is Cairo, the capital, and the "local" is the other regions. There is an epistemological distinction between the centre and the local that has been constructed in the thinking of researchers, in whose thinking the centre and the local overlap in the concepts of modernity and universality on the one hand, and the concepts of tradition and particularity on the other.

Therefore, it is impossible to observe the interface between the centre and the local, presupposing from a systematic viewpoint that a qualitative difference exists between them. To avoid this epistemological trap, it is necessary to deconstruct the relationship between the centre and the local through the

1 The Western Desert, which borders on Libya, has also been called the Libyan Desert. The governorate of the Western Desert is named Wadi Jadid, which means New Valley, in comparison with the Nile Valley. The Nile Valley signifies the Nile basin, including Lower and Upper Egypt.

 | DOI 10.1163/9789004317390_002

continuous practice of feedback, sometimes reflecting the centre in the local and sometimes reflecting the local in the centre.

Certainly, much has been discussed with respect to the local. However, it is possible to deconstruct the local, unless the discussion relates to the centre without deconstruction. The only effort required is to project the centre into the local and to look for the influences of the centre in the local without reflection.

The deconstruction of the local follows the opposite process—that is, to project the local into the centre. The point of reference for this process is to establish the local as a centre. In this manner, the centre could truly be deconstructed, and the relationship between the centre and the local would appear not as concentric circles with only quantitative differences but as a web of many relationships that are different in quality and that are segmented and overlapping, with a number of centres and localities.

Egypt has usually been described as a typical agrarian society, characterized by the centralized management of water, large scale land reclamation, labour intensive agriculture, systematic crop rotation and so on. However, this is an image that was created after the July Revolution in 1952 when the population rapidly increased, an image emphasized by the revolutionary regime to promote the national identity as an agricultural society.

This image tends to turn our eyes away from the historical transformation and the variety of agrarian life in Egypt through history. In fact, the social structure of rural Egypt today was formed in the process of modernization and colonization in modern times, from the beginning of the 19th century. The actual social structure and the spatial configuration of Egypt have been less centralized and homogeneous than is claimed by this image.

Doubtless, if Egypt were viewed from a local perspective, the locality would be taken as the centre. For example, since the middle ages, the region of Upper Egypt was the land of escape for political forces that had been defeated in the power politics of Cairo, when the forces resolved to strive for victory with redoubled efforts. In this context, Upper Egypt was the local space in relation to the centre.

However, Upper Egypt was also the space in which the original culture peculiar to it persisted. Its heroic ballads of Arab nomads remain well known to the public today. In fact, it was closely linked with the Arabian Peninsula by caravan routes in the 19th century. At the same time, the caravan routes were roads of pilgrimage to Mecca.[2]

2 The most famous is the ballad of Bani Hilal, who migrated from Hijaz and Najd in the Arabian Peninsula to North Africa, including Egypt.

In the middle of the 19th century, the Islamic doctrine of the Saviour (Mahdi), which was a traditional belief in the Islamic revolution, was expanded in Upper Egypt. Religious people who called themselves the Mahdis, which literally means "rightly guided", were almost all immigrants from the Hijaz region of the Arabian Peninsula, where the two holy places of Mecca and Medina are located, to Upper Egypt, where they settled (Barakat 1991).

The inhabitants of Upper Egypt called themselves the Arabs (*ʿurban*) and looked down upon the inhabitants of Lower Egypt as peasants (*fellahin*). In this situation, with the Arabs being confronted by the Egyptians, the position was reversed in favour of the local.

The reversal happened dramatically, when Egypt was redefined from the "local" of the desert regions outside the Nile Valley. As mentioned above, the image of the Egyptian nation-state was formed around the concept of a homeland in the Nile Valley because since the old days, Egypt had been a typical hydraulic society, totally dependent on the water of the Nile.

From the viewpoint of those in the desert, which accounted for approximately 95% of the total land area of Egypt, the Nile Valley was a huge oasis. In fact, the Bedouins who lived in the desert called the Nile Valley the "country" (*rif*), meaning the village or rural area. Those who dwelt in it were the peasants.

In reality, this was the class that was exploited by the state and the wealthy class in towns. However, they had been exalted as a symbol of Egyptian national identity in the process of nation-state formation since the 19th century. The nation of Egypt was first of all the peasants.

Conversely, the Bedouins who dwelt in the desert were in sharp contrast with the peasants of the villages. In fact, at the end of 19th century, the Bedouins were humiliated, being regarded as a sort of marginal vagabond (*tasharrud*), without fixed residences or jobs. Along with gangs, robbers, and gypsies, they were excluded from the nation of Egypt.[3]

The longing of the Bedouins for the Nile Valley was apparently projected in their calling it the "country". Their longing was for the wealth of the Nile Valley, where stability, food security, and a comfortable material life were obtainable. Nevertheless, as in their spiritual life, they had their own tastes and interests.

3 The Bedouin were not totally incorporated into the Egyptian nation on a formal level, in that they had a privilege of exemption from military service that was not permitted to other Egyptians until 1949, when the privilege of exemption from draft, granted to the Bedouin tribes, was withdrawn. Finally, in 1956 after the July Revolution, all administrative laws and regulations applied to the Bedouin tribes living in Egypt (Kato 2001:179–180).

The Bedouins had a strong love for, and pride in, their own culture. Daily life in the desert far from the centre gave rise to the original culture of the Bedouins.[4]

A Society Based on Uncertain Conditions

The relationship between the centre and the local in the context of Egyptian historiography is the hierarchical relation between rule and dependence, because the qualitative difference in spatial terms can be reduced to a quantitative difference. Indeed, studies of Egypt generally use only one regional distinction, that between Lower and Upper Egypt. Yet the adequacy of these regional categories as socio-economic categories has not been examined. Explanations of regional differences have been limited to political and cultural ones.

However, as pointed out above, actual historical and contemporary Egyptian society is spatially diverse in cultural and socio-economic terms. The image of Egypt as a centralized and homogeneous society has been challenged, especially when attention is paid to the oasis region far from the Nile Valley, where life is totally dependent on underground water.

The life of Egypt, which is symbolized in the expression "Egypt is the gift of the Nile", is predicated on the sustainability of the water supply from the Nile. Certainly, the volume of water supplied by the Nile has fluctuated and is now unable to meet the needs of the increasing population of Egypt. However, the Nile has supplied water every year and is supposed to supply it "eternally". Conversely, the ground-water is unrecoverable. Thus the society dependent on ground-water faces the possibility of sudden discontinuation because of the "uncertainty" of water resources.

This book is a community study of an oasis village called Rashda, in the Dakhla Oasis in the Western Desert. It is intended to be an area study and to provide a critical examination of the simplistic view of Egypt as a homogeneous society, especially the centre–local dichotomy. In the oasis region, a social structure and a culture different from those in the Nile Valley have developed.

However, the scope of the book goes beyond an area study. For a long time, the limits of the Earth's resources have caused alarm, and "uncertainty" is the key word in understanding the future of human beings. In this regard, this book presents an interesting case study of a way of life in a society that is

4 The lifestyle peculiar to them and their affection for it caused widespread resistance against conscription for military service among settled Bedouins and provoked a large-scale revolt against the state led by Umar Masri, Shaykh of the Jawazi tribe, in the middle of the 19th century (Kato 1990, 2001). On the Jawazi tribe in Egypt, see (Kilani 1994).

strictly conditioned by marginal and uncertain circumstances and that is dependent on limited water resources.

The three key concepts for the discussion are: the uncertainty of the water supply, dependence on the political regime, and the rational behaviour of individuals.

1 Uncertainty of the Water Supply

The availability of water determines economic life based on agriculture in an oasis village such as Rashda. Water is an essential factor for agriculture in the Nile Valley as well. However, the latter depends on the Nile, and so the future availability of water can be controlled and estimated. In contrast, in the oasis region where agriculture is totally dependent on ground-water, it is hard to plan the use and management of water in the medium and long term, because the ground-water flow is uncontrollable. Hence, a direct relationship between the location of wells and springs and the expansion of cultivated lands is observed. In fact, the farmers' interest lies in the quantity of water, rather than in the size of their plots.

2 Dependence on the Political Regime

The unsustainability of water is a characteristic peculiar to an oasis economy. Therefore, the survival of the oasis economy largely depends on the financial and technical conditions at any given time. In modern times, the circumstances of water affairs have become severe, because of the necessity for digging deep wells and obtaining sufficient water. The increasing population requires large amounts of capital and technology. This demonstrates the involvement of the state and private enterprises in irrigation agriculture in an oasis region. In fact, the use of water in agriculture leads to a decline in the underground water level. Therefore, digging wells to deeper levels is necessary to maintain access to it. This in turn requires a large amount of money. Thus, although the oasis region may appear to be on the periphery, its survival largely depends on the general political and economic regime.

3 Rational Behaviour of Individuals

The farmers have developed a cultivation system to adapt to natural conditions characterized by the scarcity of water resources. This point is related to the local cultivation customs in the village. The farmers cultivate in a way that they have developed following their experience. For example, the way in which water is drained has an important effect on agricultural irrigation and is necessary to avoid soil becoming too saline. Yet soil salinization is a phenomenon that appears here and there in the irrigated areas of the village. However, this does not mean that the farmers are not conscious of the problem of salinization. Rather, the farmers consider the problem of soil salinization in relation to expenditure. They consider the quality of the soil, the availability of capital, and the quantity of water, so as to select the most efficient method of irrigation.

Methodology and Data

Methodology

A multi-perspective approach is adopted in this study. The multiple perspectives signify the combination of different disciplinary tools of analysis, a communal approach, and methodological individualism.

As mentioned above, Egyptian society has been viewed as a hydraulic society. The result of such a view has been an emphasis of the normative aspect of the social structure, which is a corollary of the authentic view of the relationship between the centre and the local, emphasizing the dichotomy of rural–urban and Cairo–other regions.

This could be called the "communal approach". Many of the studies taking this approachhave focused on factors such as kinship and social customs to determine the nature of society.[5] "Methodological individualism", on the other hand, is the method commonly employed in the social sciences. It emphasizes the rational behaviour of the individual.

These two approaches are compatible. They interact and supplement each other, and determine the orientation of societal development. For this purpose, it is necessary to collect the empirical data and information at the individual or household level by conducting social surveys. It had been difficult to do this, mainly because of military and security restrictions. Fortunately, however, these constraints on research were relaxed after the introduction of the open economic policy in the 1980s. Surveys in urban and rural areas are now becoming easier, and the environment is better for the examination of empirical studies using many different materials and methods.

Data

A multi-perspective approach was made possible by the use of various kinds of data and information. This book is based on information from the following sources.

5 On the "communal approach" in Japan and France, see Kato 2008, Iwasaki 2006, 2007.

1. Agglomerated statistical data at the lowest administrative level of the village (*qarya*), produced by the Egyptian government.[6]
2. Micro-level data collected through household surveys.
3. Geographical information, including village maps with details at the level of individual buildings, and various other kinds of maps such as satellite maps.
4. Historical documents possessed by village notables.
5. Other qualitative information, such as family histories and interviews with village authorities, notables, and other residents.

Of these, 2, 3, 4, and 5 were collected by field surveys in Rashda. The main sources of information gathered by field surveys were 2, 3, and 4.

1 Household Survey in Rural Egypt

The household survey in Rashda was conducted as part of a joint research project between Hitotsubashi University and CAPMAS in Egypt, which has been running since 2002. This research project aims to collect original micro data on Egyptian towns and villages; for rural Egypt, the household surveys have been conducted in 19 villages (Map 2).

The objective of the household survey in villages was to clarify the regional diversity of rural Egypt at the village level, the smallest administrative unit. The villages were selected from among the three regions: Lower Egypt, Upper Egypt, and the oasis region, after thorough discussion and on the advice of CAPMAS staff.[7]

---

6 Egyptian administration is divided into three hierarchal divisions: governorate, district (*qism/markaz*), and town (*shiyakha*) for urban areas or village (*qarya*) for rural areas. Two kinds of statistical data at the town and village level are used in this book. These data became available with the help of CAPMAS (Central Agency for Public Mobilization and Statistics) staff as part of a joint research project between the Graduate School of Economics at Hitotsubashi University and CAPMAS. An Agreement between CAPMAS and Hitotsubashi University for the Promotion of Cooperative Research and Exchange of Personnel was concluded in 2004.

7 Additional small "sampling surveys" were conducted to collect supplementary information. These were conducted before the implementation of the household survey to establish the list of households residing in the village, which was used as a basis for selecting the sample households. The information in this list included the names of all householders residing in the village, as well as their location: the block, building, and floor in which each household resided. "Sampling surveys" were continued after the household survey in 2006, to collect additional information on the year of construction of each flat in the building in which each household resided.

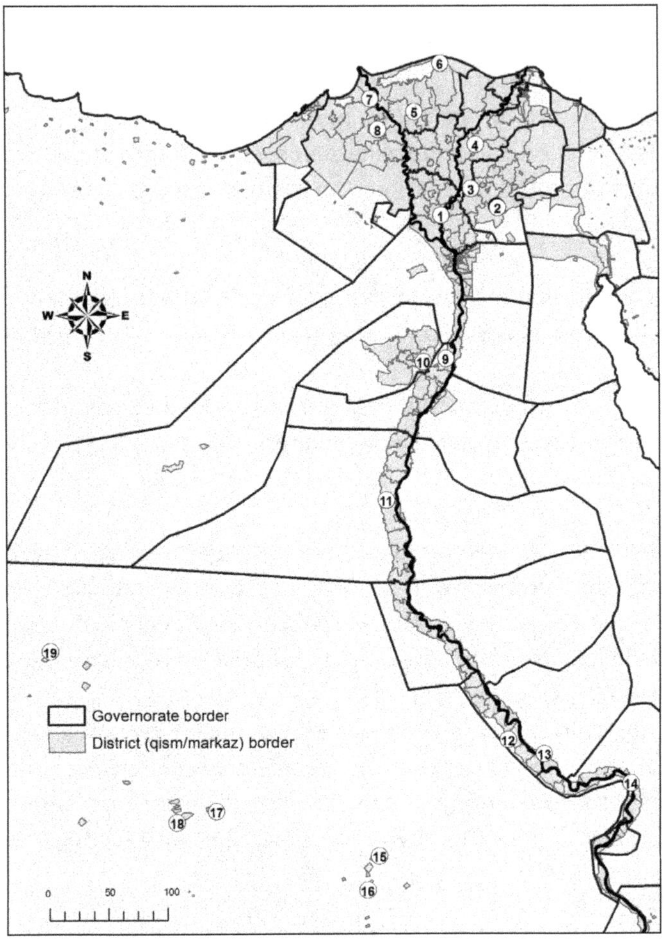

	Village	Governorate
①	Abu Senita	Menufiya
②	Zahra	Sharqiya
③	Abu Tawala	Sharqiya
④	Kafr Shubrahur	Daqahliya
⑤	Kafr Beni Hilal	Buheyra
⑥	Borr Bahry	Kafr Sheykh
⑦	Hamra	Kafr Sheykh
⑧	Sidi Oqba	Buheyra
⑨	Homa	Beni Suef
⑩	Damashqin	Fayyum
⑪	Beni Samrag	Minya
⑫	Kawamel Bahry	Suhag
⑬	Awlad Sheykh	Suhag
⑭	Sheykh Isa	Qena
⑮	Munira	New Valley
⑯	Bulaq	New Valley
⑰	Balat	New Valley
⑱	Rashda	New Valley
⑲	Liwa Subih	New Valley

MAP 2 *Location of survey villages in Egypt*

The survey questionnaire consists of the following 12 sections on a series of topics that integrate monetary and nonmonetary measures of household welfare and a variety of household behavioural characteristics.[8]

2 Making a Digital Village Map

The most important task in recording geographical information is to make a digital village map. The administrative digital maps of the smallest administrative units—towns in urban areas and villages in rural areas—were provided by CAPMAS to combine statistical data with geographical information gathered by GIS methods. However, digital maps at the building level in residential areas, which are necessary to link household survey data with geographical information, were not available. In this situation, detailed residential maps of the survey areas had to be made by the authors with the support of local informants and engineers.

To make detailed residential maps of the survey villages, 1/2500 scale paper maps, which are used as base maps, were collected from the Egyptian General Survey Authority. Then, locational information for each household was collected during the fieldwork. The information included the street name, block, building and floor of the apartment where each household resided. This information was digitized on a 1/2500 scale map that was scanned and georeferenced, with drawings of polygons to represent streets, buildings and blocks, and then the household ID, building ID and block ID were inserted into each polygon.

3 Local Documents

As mentioned above, Egypt is famous for being a centralized society. Cairo is an excellent political, economic, and cultural centre, and this characteristic of Egyptian society is typically reflected in the scarcity of local documents. This is certainly true of the Nile Valley. However, the situation is different in the oasis region, which is far from Cairo. There, a considerable number of documents are in the possession of village notables. Most of these are historical docu-

8 The 12 sections are as follows. I Basic information. II Wage employment. III Unemployment. IV Non-agricultural self-employment. V Migration (V-1 Out-migration from the village, V-2 Labour migration outside the village, V-3 In-migration to the village). VI Agricultural management (VI-1 Cultivated land, VI-2 Rental land, VI-3 Livestock, VI-4 Agricultural production, VI-5 Agricultural equipment, VI-6 Agricultural inputs, VI-7 Labour management, VI-8 Irrigation). VII Household income and expenditure (VII-1 Household income, VII-2 Household expenditure, VII-3 Transfer from/to the other households, VII-4 Credit, VII-5 Financial assets). VIII Housing conditions. IX Attitude questions. X Social relations in the village. XI Communal activities in the village. XII Parents of the head of household.

ments, but some are "living" documents to which villagers refer when necessary.

These documents are precious as source materials, being a basis on which Egyptian society can be reconsidered from a peripheral perspective. The local documents in the oasis region are now scattered to the fringes, because of the social transformation caused by the development policy of the Egyptian Government. Part of the purpose of the field survey project in oasis villages was to collect local documents and to prevent their dispersal.

Why Rashda?

The main subject of this book is an examination of aspects of a village named Rashda from multiple perspectives. Rashda was selected as a research site for the following reasons.

In the household surveys for villages for the above-mentioned research project on the social economy of Egypt, the villages were selected from those with a population of approximately 5,000, so that the sample households (600 in each village) are representative of that village.

For the household survey of the oasis region, five villages in the New Valley were selected: one in Farafra Oasis, two in Kharga Oasis, and two in Dakhla Oasis. The main criteria for their selection were first that there should be over 4,000 inhabitants, and second, its history—the villages were to be selected from both old and new villages.

Villages in the oasis region were selected, first of all, to represent the regional diversity of Egypt. The natural conditions in the oasis region are completely different from those in the Nile Valley. The villages in the oasis region are totally dependent on underground water, while the villages in the Nile Valley rely on the water of the Nile.

Secondly, villages in the oasis region were picked because they are suitable for examination of the conditions of sustainability for societies with restricted resources. The three key concepts for the above-mentioned discussion were established to address this concern.

Among the five villages in the oasis region, Rashda, the most populous village in Dakhla Oasis, attracted our attention for the following two reasons. First, it is a "new" village established after the 19th century, and we can follow the process of its birth and development—how people established their homes and their lives, and explored the underground water resources, and how they formed their social relationships in the village that they moved into.

Second, we were able to collect and use various kinds of source materials mentioned above. We have had the precious chance to conduct field surveys with the support of a talented informant. In addition, Rashda still has old documents that date back more than 100 years, which makes it a very rare case.[9]

In conclusion, it is an experiment in both subject and method to write this book. Theoretically, as mentioned above, we adopted a multiple perspective, using various kinds of data and information. Regarding the subject, we attempt to describe the history of modern Egypt from the viewpoint not of the centre but of the periphery. This viewpoint is very useful, because we believe the paradoxical phenomenon that the spirit of the time is more critically reflected in the periphery than the centre.

Framework of the Book

This book aims to provide a comprehensive description of Rashda, a village in Dakhla Oasis, as a society based on underground water, from its beginnings to the present. As mentioned before, key concepts are the uncertainty of the water supply, the dependence on the political regime and the rational behaviour of individuals. The villagers of Rashda have created local customs of irrigation and cultivation under the circumstances of the uncertainty of the water supply, and its development depends ever more on the political regime, as long as large amounts of finance and superior technology are necessary to dig deeper wells to secure water for cultivation.

The book is composed of four parts.

Part 1 deals with the region of Dakhla Oasis to examine the natural and human circumstances of an oasis village like Rashda. It points out the main features of Rashda and compares them with other regions of Egypt by locating the region of Dakhla Oasis in Egyptian society from the viewpoints of periphery in the local administrative system (Chapter 1), the conditions of natural and human resources, that are water and population (Chapter 2), household and family structures (Chapter 3), and the social economy, especially income level, the occupational situation and educational standards (Chapter 4).

Part 2 reconstructs the history of Rashda, a new village formed in the 19th century, from the end of the 18th century until the present, mainly using local documents. The history of Rashda can be divided into three stages: Rashda as a field (down to the beginning of the 19th century, Rashda as a settlement in the middle and latter half of the 19th century (Chapter 5), and Rashda as an independent village from the end of the nineteenth century (Chapter 6). From

9 On this point, the works of Rudolph Peters (Peters 2007, 2011) are epoch-making, because they are based on the local documents possessed by village notables.

this historical perspective, Rashda is described as a society based on water (Chapter 7).

Part 3 makes a full examination of the features of Rashda today as a community, mainly using the statistical data on households and geographical information collected during field surveys. Rashda as a community is formed based on official institutions and so-called unofficial customs. The first is concerned with its public functions as an administrative unit (Chapter 8), and the second with the social relations based on the structures of household and family (Chapter 9), the organization of the digging and management of wells and springs (Chapter 10), and the cultivation system of agriculture in the household and market economies (Chapter 11).

Part 4 deals with an Irrigation District called "Well No. 3 Irrigation District" as a case study to discuss Rashda as a community, using statistical and geographical information, including digital maps, satellite data, and the other qualitative information gathered during field surveys. After defining the term "irrigation district" assigned to a specific well or spring (Chapter 12), the features of "Well No. 3 Irrigation District" are examined from the viewpoint of water management and the cultivation system (Chapter 13) and social relations in the District (Chapter 14). This District was selected because it was considered to be a good community study to examine the sustainability of a society living under conditions of limited water, with reference to the relationship of the three key concepts: the uncertainty of the water supply, dependence on the political regime, and the rational behavior of individuals.

PART 1

Dakhla Oasis

CHAPTER 1

Islands of the Blessed

Rashda is located in Dakhla Oasis. Dakhla literally means "inside" in Arabic. As is suggested by its name, Dakhla Oasis is situated deep within the Western Desert in Egypt. In the pre-modern age, the existence of Dakhla Oasis was known in literature, but descriptions of it are almost just the images by a kind of imagination. As for the history of Dakhla Oasis in modern times, it was the process of its integration into the nation state of Egypt that began to be formed under the reign of Muhammad Ali (ruled 1805–48). This chapter describes its process particularly from the viewpoint of the vicissitude of local administration.

Islands in the Sand

Historiography on Dakhla Oasis

The Western (Libyan) Desert forms the eastern part of the Sahara Desert.[1] The Western Desert covers an area of 1,600 km from near Kordofan in Sudan, extending north through Egypt to the Mediterranean coast. It includes all the land of Egypt west of the Nile, covering two-thirds of the entire country, more than 680,000 square kilometres (Sampsell 2003: 137).

Beginning at the Nile Valley, it stretches west for 1,760 km. The first 500 km are within Egypt's present-day borders, the desert thereafter extending far into Libya. It is also the most arid region on earth; 58°C (136 °F) was recorded south of Tripoli in Azizia on September 13, 1922, which is the highest temperature ever recorded (Vivian 2008: 4).

In spite of severe natural conditions, since ancient times, the oases have been inhabited by people who made use of the artesian springs at the lowest parts of the depressions in this desert. Free-flowing wells fed by the Nubian aquifer system underlying the major oasis make irrigated agriculture possible.

Nowadays, there are five known oases or depressions in the Western Desert: the Bahariya, Farafra, Kharga, Dakhla, and Siwa Oases.[2] They are administered

1 According to Vivian, the name of the Western Desert was invented by the British to label their share of the North African wasteland. "Mountains separate the Libyan Desert from the Sahara Desert, but no natural barrier distinguishes the Western Desert from the Libyan Desert. The Western Desert exists only as a political demarcation" (Vivian 2008: 1).

2 Kharga and Dakhla mean "outer" and "inner" in Arabic.

 | DOI 10.1163/9789004317390_003

by the following governorates: Bahariya Oasis belongs to the Giza governorate, Siwa Oasis belongs to the Matruh governorate, and Kharga, Dakhla, and Farafra Oases belong to the New Valley governorate (see Map 1).[3]

It takes approximately 10 hours to travel by car from Cairo, the capital of Egypt, to Dakhla Oasis. It is located 750 kilometres south-west of Cairo in the Western Desert. There are many historical remains of the Egyptian and Roman civilizations, which are now buried by sand. These, which appear so impressive in the sunset, remind us that life in the oasis has continued under very unstable natural conditions.

Dakhla Oasis was known in ancient times, as shown by many historical remains. Many people stayed and lived there in settlements, some of which were prosperous. The following sentences on Kharga and Dakhla Oases are taken from the famous book entitled *History* by Herodotus, the Greek historian of the 5th century BC.

> So fared the expedition against Ethiopia. As for those who were sent to march against the Ammonians, they set out and journeyed from (Egyptian) Thebes with guides; and it is known that they came to the city of Oasis, inhabited by Samians said to be of the Aeschrionian tribe, seven days' march from Thebes across sandy desert; this place is called, in the Greek language, the Islands of the Blessed (Herodotus, Book 3 Chapter 6 Section 1).

Oasis was the name given to the oases in Egypt by the people of the ancient Mediterranean.[4] The word "oasis" comes from Greek and is originally from the Arabic word "wah". "Wah" probably meant "dwelling place" in Demotic Egyptian.

For the ancient Greeks, the oases were those to the west of the Nile, which formed the Egyptian dominion between the Nile and Libya. Herodotus seemed to know only one, which was Dakhla Oasis. Ptolemy knew two of them, Kharga

---

3 Currently, the governorate division is under study. The new demarcation of provincial borders will be conducted in three stages. The first entails remapping the boundaries of the southern Upper Egypt governorates. The second encompasses Cairo, Giza, Suez, and the southern and northern Sinai governorates, and the third will see the birth of three new governorates: Central Sinai in the Sinai Peninsula, Alamein in northern Matruh, and Oasis (Wahat). According to the plan, the oases in the Western Desert will be divided into three governorates: Siwa will be under the Alamein governorate, Bahariya and Farafra under the Oasis governorate, and Kharga and Dakhla under the New Valley governorate.

4 Balat in particular, according to Garcier and Bravard (2014: 1).

and Dakhla Oases, and Strabo knew three oases; namely Small Oasis, Great Oasis, and the Oasis of Ammon (actually Siwa Oasis).

Geographers and historians in the Islamic period also knew the oases in the Libyan Desert. al-Masudi (896–956) is known to be the first Arab writer to refer to the oases in the Libyan Desert. He states in his book *Meadows of Gold and Mines of Precious Stones*, "The land of oases is that of palm leaves, wonders and acid water".[5] He regarded the Libyan Desert as an independent area of Egypt.

After al-Masudi, several historians and geographers, such as Ibn Duqmaq (d. 1406), Ibn Hawqal (d. 977), al-Idrisi (1100–1166), Ibn Mammati (d.1209), Ibn al-Wardi (d. 1349), and al-Maqrizi (1364–1442), referred to the Libyan Desert. However, their descriptions concerning the position and extent of the oases in the Desert are vague, and the oases may have had so-called imaginary boundaries, except on the side towards the Nile (Rennel 1830:552).

According to Rennel, these Arab geographers seem to describe the oases by the collective term "wahat", the desert that contains the "wahs" or oases dependent on Egypt. For instance, Abu al-Fida' (1273–1331) wrote of a "wahat" composed of a number of tracts that were like islands surrounded by the Desert, and he mentioned that there was a space of three journeys between it and Said (Upper Egypt) (Rennel 1830:190, 191)

It seems that the oases in the Western Desert lost their significance during the Islamic period. For example, al-Idrisi wrote "Al Vahat is near Assouan, to the west of it. This region, which is now desolate and uninhabited, was formerly well peopled. Streams run through it, and trees and deserted towns still remain. Therefore, from the rear of this region to Cavar and Cucu, the country is covered with palms and ruins of buildings" (quoted in Edmonstone 1822: 118).

The oases began to attract attention from the 19th century during the era of "scientific exploration". Since the 19th century, many European explorers have visited Kharga and Dakhla Oases, such as Archibald Edmonstone with a scientific ambition of discovering the ancient civilization in the desert.[6] This ambition of European explorers coincided with Muhammad Ali (who ruled from 1805 to 1848) and the Egyptian government's ambition to establish a nation-state.

In modern times, besides the scientific concerns mentioned above, particular attention has been paid to the oases in the Western Desert in terms of the administrative concerns of the Egyptian government for transit nodes for long-distance trade between Sudan and the Nile Valley, and between Libya and the

5 See the English translation of al-Masudi' text in Lunde and Stone (1989).

6 See the Glossary of Travellers and Explorers (Vivian 2008: 421–439).

Nile Valley on the one hand (Walz 1978), and the development of the desert for agriculture on the other.

Central governmental control had not yet been achieved in the oases in the Western Desert in the 19th century, as shown in the large-scale revolt by Bedouins in the middle of the 19th century.[7] The settlement policy for the Bedouins could not be achieved until the first half of the 20th century. The process of expansion of the central governmental control in Dakhla Oasis was that of the centralization of administration, which will be mentioned below.

Islands in the Sand

Greek historians and geographers such as Herodotus and Strabo compared an oasis to an "island in a sea of sand". In fact, the similarity of an oasis to an island is suggestive. Keiichi Takeuchi defines an island as any area of land smaller than a continent and entirely surrounded by water, and pointed out eight attributes of islands as follows.[8]

1. Isolation
2. Offshore advantages
3. Powerlessness or peripherality
4. No universal panacea for island economy
5. Ethnic plurality
6. Geopolitical significance
7. Ecology
8. Geographical imaginations of the island

These attributes seem to be applicable not only to an island but also to an oasis, if water were replaced by sand. Ellen Morris discusses the same idea, referring specifically to oases in Egypt, citing the words on Dakhla Oasis by Herodotus mentioned earlier.

---

7 The revolt of Umar Masri was a widespread rebellion with an impact that extended to Asyut Province in Upper Egypt and the oasis regions of the Libyan Desert, as well as the Minya and Fayyum Provinces in Middle Egypt. In addition, other major disturbances were caused by Bedouins in the reign of Said Pasha (who ruled from 1854 to 1863) not only in Upper Egypt but also in Sharqiya and Qalyubiya Districts in Lower Egypt, for which there is ample documentary information in the Egyptian National Archives. It shows explicitly that the Bedouin disturbances spread over almost all of Egypt in the middle of the 19th century (Kato 2001:182–84).

8 Keiichi Takeuchi, "The Insularity in the Contemporary World—Small Islands in Japan, Italy and the World", the report of the meeting of the Mediterranean Studies Group, Hitotsubashi University, on 21 November 2004 <http://www2.econ.hit-u.ac.jp/~areastd/mediterranean/download/pdf/The_Insularity2.pdf>.

> In the Aegean isles, the ease of marine transport, the proximity of other islands, and a wealth of natural resources, ensured that no island remained an island unto itself, as it were. These islands, since the advent of boatbuilding, were defined as much by connectivity as they were by insularity. But what about oases? Might the isolated oases of Egypt's Western Desert actually constitute a richer more interesting subject of meditation for "island theorists" than Mediterranean islands? The likening of oases to islands is certainly not a new idea, for Dakhleh and Kharga were known in Herodotus' writing as the "Islands of the Blessed" (Herodotus 3.26.1). Strabo too stated that "the Egyptians call the inhabited tracts, surrounded in a circle by great deserts, like islands in the sea, Auases." (Morris 2010: 127–128)

In summary, villages at an oasis could be called a society of "connectivity". Here, "connectivity" refers to the various relationships between villages at the oasis and the surrounding communities, which are not capable of self-sustenance and need to trade with each other for their existence.[9] As a result, migration is a typical phenomenon of oasis villages in the uncertain circumstances of water resources.

9 The word "connectivity" was borrowed from P. Horden and N. Purcell. They were inspired by a landmark book on Mediterranean studies by F. Braudel, *The Mediterranean and the Mediterranean World in the Age of Philip II* (1949). They attempted to demonstrate in their book *The Corrupting Sea: A Study of Mediterranean History* (2000) that the Braudel thesis of the Mediterranean world as a historical entity could be applied not only to the 16th century but also to the periods in history from the classical to the modern age. The discussion of Horden and Purcell is based on two analytical concepts: "micro ecology" and "connectivity". "Micro ecology" refers to societies with different ecological circumstances, and "connectivity" refers to the various relationships between them. In their opinion, none of the societies surrounding the Mediterranean Sea is capable of self-sustenance, and they need to trade with each other to survive. This means that the Mediterranean world is one of "connectivity". Thus, the societies around the Mediterranean Sea cannot stay pure, and this is why the Mediterranean is called the "corrupting sea", a sea that corrupts the purity of its societies. In this sense, the Western Desert of Egypt is also a world of "connectivity", where the components of oasis villages are incapable of self-sustenance and need to trade with each other and with the outer world to survive.

Dakhla Oasis in the Administration of Modern Egypt

The history of Dakhla Oasis in modern times was the process of its integration into the centralized administrative system centred on Cairo. It is explicitly reflected in the changes of local administration at Dakhla Oasis.

Dual System of Hukm and Qada'

Before the introduction of modern judicial systems at the end of the 19th century and the beginning of the 20th century, villages at Dakhla Oasis were governed under the dual system of the *hukm* and the *qada'*. The *hukm* meant the administrative hierarchy, at the top of which was the *hakim* (governor), and the *qada'* signified the judicial hierarchy, at the top of which was the *qadi* (judge) of the *mahkama* (religious court).

This dual system can be confirmed until the middle of the 19th century, based on local documents.[10] Then, Dakhla Oasis came under the *hukm* of the Asyut governorate (*mudiriya*) and under two *qadi*. This meant the abolition of the traditional administrative system of the *hukm*, and Dakhla Oasis was integrated into the centralized administrative system.

In 1292 AH (1875), the *qada'* was assumed by the *qadi* of *mahkama*. After his death in 1298 AH (1881), Ahmad Yusef Muharram was appointed as the *ma'dhun* of Qalamun. This entailed a change in judicial system from the traditional *qada'* system to the *ma'dhun* system. In Islamic law, the title of the *ma'dhun* (*al-ma'dhun al-shar'i*) refers to the official authorized by the *~~al-~~qadi* to perform civil marriages.[11]

However, in the local context of the Dakhla Oasis, he also had a role as the secretary of the religious court (*qalamiya al-mahkama*) to draw up official documents not only for civil marriages but also for civilian affairs such as water distribution. The *ma'dhun* was not appointed by the authorities but was selected according to local customs (*'urf*). The post of the *ma'dhun* has usually been assumed by members of prominent families

Introduction of the Umda System

The foundation of the post of the umda (*'umda*) is a substantial turning point in the history of the local administration of modern Egypt. In spite of the meaning of the word in Arabic, the umda is not a village representative but

10 According to the informant, the description in the document was based on 20 local documents without the writer's name or a date.

11 The importance of the *ma'dhun* in local politics is certificated by the frequent reference to them in local documents and the existence of the successive *ma'dhun*.

rather a village official appointed by the state. Historically, this post was founded in the process of rural reorganization for the purpose of administrative centralization in modern Egypt in the reign of Muhammad Ali in the 1830s (Kato 1980).

Before this rural reorganization, and under the dual system of the *hukm* and the *qada'*, the traditional village official was called the *shaykh al-balad* (village shaykh), who managed the village administration. He was the representative of the *hissa* (village quarter) as well as of the largest family (*'a'ila*) who lived in the *hissa* concerned. Therefore, the *hissa* was named by the chief of its largest family, and several *shaykh* were appointed in one village for the village administration, in cases where the village had several *hissa*.

In contrast, one umda was appointed in every village. While the *shaykh al-balad* was the traditional leader of village society and represented the villagers to the state, the umda was literally responsible for village control under the direct supervision of the state, although both of them were village officials standing between the state and village society.[12]

The umda was a so-called "modern local administrator" who could manage the formal and governmental documentations concerning his village. Village politics were developed around the post of the umda; many disputes arose because of abuses of power by the umda.[13]

The post of the umda was first introduced in Qalamun in 1299 (1880–1881) in Dakhla Oasis. It is significant on this point that since then, village inhabitants have not been permitted to dig a well without the permission of the authorities.

In 1882, Dakhla Oasis belonged to Asyut governorate as Dakhla Oasis district (*markaz*).[14] The administrative units to which Dakhla Oasis has belonged since then were as follows: Asyut province from 1897 to 1907, East Desert and Oases province in 1917, Southern Desert governorate from 1917 to 1947, and the New Valley governorate from 1960.

During these periods, the villages of Dakhla Oasis evolved. In 1882, there were four villages in Kharga Oasis and 10 in Dakhla Oasis.[15] Among these 10

12 Recently the influence of the umda has been decreasing in village politics. On this subject, see Chapter 8.

13 Illegal military conscription is treated as a typical abuse of power by the umda (Kato 1995).

14 It was called "Wahat Asyut" (Dakhla and Kharga Oases).

15 In the population census of 1882 and 1897, a village is called a *nahiya* (plural *nawahi*). In 1882, Kharga Oasis had the four villages of Kharga, Baris, Bulaq, and Gunah, and Dakhla Oasis had the 10 villages of Ismant, Balat, Budkhulu, Jadida, Hindawi, Qalamun, Qasr, Maasara, Mut, and Mushiya.

villages in Dakhla, Qasr village had five dependencies and Qalamun village had four, among which was Rashda.

The number of villages listed in the 1897 population census increased to 12, because Rashda was listed separately from Qalamun village as Nazla Rashda, and Tenida was listed separately from Balat village as Nazla Tenida. The number of villages remained the same until 1947.

In 1966, some new villages were created, such as Baghdad, Munira (previously Mahareq), Nasir al-Thawra, and Sanaa at Kharga Oasis but not Dakhla Oasis. In 2006, many new villages were created during the period 1996–2006, but most of them were settlements that belonged to other settlements. Thus, other than some villages during the 1950s to 1960s, no new villages were created there.

Introduction of the Local Unit System

The next turning point in the history of the local administration in Dakhla Oasis was in the 1960s under the socialist regime of Gamal Abdel Nasser. From 1960 to 1965, the village administration was reorganized under governmental direction with the establishment of the New Valley governorate.

In 1963, a committee of the socialist union was established in every village. The committee succeeded in forming a Local Unit (*al-wahda al-mahalliya*), which was responsible for the execution of governmental policies, and a Local People's Council (*majlis al-sha'abi al-mahalli*). This was elected by direct secret ballot for a four-year term and was responsible for supervising the performance of a local unit established under Law 124 of 1960.[16]

Legally, all the local affairs in social, economic, and health matters are in the hands of a local unit. The first article of the local law No. 43/1979, created by law No. 145/1988, stipulates that the village local unit is an administrative unit at the village level (DPADM 2004: 7). It is also a governmental organization charged with bringing about vital changes in the functions and structures of the village social systems.

The local unit is to establish and manage all public infrastructure located in its geographical area. The village local unit conducts its specified tasks through

16 According to the Constitution of 2014, every local unit shall elect a local council by direct and secret ballot for a term of four years. A candidate shall be at least twenty-one (21) Gregorian years of age. The law shall regulate the other conditions for candidacy and procedures of election, provided that one-quarter of the seats shall be allocated to youths under thirty-five (35) years of age, and one-quarter shall be allocated to women, and that workers and farmers shall be represented by no less than 50 per cent of the total number of seats, and these percentages shall include appropriate representation of Christians and people with disabilities (State Information Service <http://www.sis.gov.eg>).

the Village Local Public Council and Local Unit, which is the executive administrative authority.[17]

An executive council of villages is formed by the heads of each village's local units and all the heads of the executive authorities within a village. They represent the Ministries of the Interior, Education, Social Affairs, Health, Agriculture, and Housing as well as the village bank. This executive council of villages takes responsibility for supervising the executive authorities, implementing the decisions and recommendations of the local people's council, and supporting the head of the local unit in tasks such as executing and financing administrative plans.

Land Exploitation in the Oasis Region

The literature on the modern history of the western oasis region points out that before and after World War II was a dark period of economic affairs.[18] This fact is reflected in the demographic trends in the 20th century.[19]

In the contemporary context, irrigation development in the Western Desert officially began with the establishment of the Desert Irrigation Circle (*taftish rayy al-sahara*) in Kharga city in 1938, after some failed private projects such as the one conducted by a private English company in the areas of Sharika and Mahariq in 1905. The circle helped the inhabitants of the oasis to dig wells and to dredge springs. The district increased its activities after 1950 by using more advanced excavators that had been used for digging oil wells (Husayn 1975: 92–93).

Then a new era began. After the July Revolution in 1952, the Permanent Organization for Agrarian Reform (POAR) extended its activities into the Western Desert in 1957. It reclaimed land using the new water resource provided by the irrigation projects of the district and distributed the reclaimed land to the inhabitants.

The land reclamation the New Valley was a part of the broad nationalistic and socialist plan to solve the food security problem and to improve rural

17 Article 66 of law No. 43/1979 and the law of local rule No 50/1981 and law 145/1988 stipulates that each unit should have a local people's council consisting of 20 members to be elected from the public leaders of the villages served by the local unit. This council has the role of planning the economic, social, and rural development of the villages in ways such as suggesting projects for public benefit and approving the final statements, suggesting methods of public participation using the residents' own efforts and competences, and promoting the establishment of public infrastructure in the villages (Mayfield 2012: 101–105; Ibrahim 2006: 20–29).

18 For example, (Vivian 2008: 125) and (Fakhry 1974: 30, 176).

19 This subject is discussed in detail in Chapter 7.

livelihood. It can be considered to be an extension of the agrarian reform that was undertaken from 1952, only two months after the July Revolution.

Egyptian land reclamation projects have undergone three main phases since 1952, influenced by the results of the project itself, and by changes in national ideology and circumstances: 1952–1959, 1960–1967, and post-1967 characterized by the retrenchment of government projects and a shift towards privately initiated large farm projects.

The first phase from 1952 to 1959 was of experimental farms and pilot projects. This phase was rather an experimental stage, characterized by a fascination with the concept of projects. The main project begun in this phase was the creation of Tahrir (Liberation) Province on the western fringe of the Nile delta, initiated by POAR in 1953. "It was envisaged not merely as an agricultural project, whose scale was ambitious in itself, but as an entire social experiment, the nucleus and vanguard of Egypt's new rural society".

Another project that was proposed in the late 1950s for the expansion of agriculture was "the creation of a New Valley, sister to the Nile, in the Western Desert". It was an ambitious plan to create a second valley in the Western Desert in addition to the old valley, the Nile Valley, so that the surplus population in the latter could move into the new reclaimed land.

After the surveys of resources by a team of geologists in 1954, 22 artesian wells were drilled in Kharga and Dakhla Oases that seemed to yield a promising amount of water. Some experimental studies were carried out and gave encouraging results. This led to the establishment of the Egyptian General Desert Development Organization (GDDO) in 1957.

The second phase was between the initiation of the First Five Year Economic Development Plan (1959/60–64/65) in 1960 and the 1967 Middle East war. The object of the plan was to reclaim 521,000 *feddan*: 466,000 *feddan* irrigated with existing Nile water supplies, 25,000 *feddan* in South Tahrir using underground water for sprinkler irrigation, and 30,000 *feddan* using water that would be available from the Aswan High Dam. The efforts of this plan were largely successful, reaching 87 per cent of target (Voll 1980:129, 131, 132–33).

GDDO started the New Valley in October 1959 within the framework of the plan (Husayn 1975: 93). The Five Year Plan proposed the reclamation and cultivation of 121,000 *feddan* in the New Valley governorate, of which 96,000 *feddan* were to be in Dakhla and Kharga Oases. This project covered a vast area of about 458,000 square kilometres extending from the Kharga, Dakhla, and Farafra oases (Vivian 2008:60). Along with the Aswan High Dam (the construction of which started in 1960), the New Valley Project constituted one of two plans implemented by the Arab Socialist regime to remodel the natural environment. The inspiration for the project was the historical accounts of thriving well-populated settlements around the oases in ancient times (Voll 1980: 130–31).

However, the reclamation of desert land faced difficulties. Soon after, preliminary surveys revealed that the original plan had overestimated the water resources. Furthermore, facing financial difficulties, the New Valley Project was obliged to revise its target for land reclamation. By 1966, the actual area that had been reclaimed was only 44,000 *feddan* (18,500 hectares). Moreover, 16,000 *feddan* (6,700 hectares) of this land, 36 per cent, were not cultivated because of water problems (Husayn 1975: 101).

The most crucial problem was the cost of water extraction. While groundwater exists for irrigation, only a small portion of the land can be irrigated economically in the New Valley. Extraction is limited not by the quantity of water but by the depth and expense of pumping. In addition, the cost of reclamation and cultivation was excessively high.

It is difficult to evaluate accurately the final performance of the New Valley Project because of a lack of accurate data, and this is also true of other land reclamation projects conducted since the revolution. It is said that a number of failures in water management (over-irrigation and salinization, in particular) and cultivation operation resulted in many areas of reclaimed land being abandoned.

However, it is undeniable that the New Valley Project was the starting point for socio-economic development in the Western Desert. Distribution of 5-*feddan* plots to impoverished or landless peasants, which was the most common method of disposing of reclaimed land in the New Valley, was considered to be an exercise in community development as well as agricultural production. Such development included transportation (for example, paved roads and Kharga Airport) and communication, basic infrastructure, immigration from the old valley, especially Upper Egypt, and industrial and commercial development. At the village level, each community was supposed to have schools and vocational training, medical facilities, mosques, youth centres, marketing areas, and administrative buildings (Voll 1980:142, 145). The impact on the development of Dakhla Oasis can be seen in Excursus 1 on the local history of Dakhla Oasis in this Chapter.

The 1967 war brought the governmental projects, including the New Valley project, to retrenchment, because of a shortage of funds. The governmental project shifted towards the redesign of the already reclaimed land for effective use, such as land sales instead of state farms or a leasehold system, and graduate programs implemented in areas such as Farafra Oasis.

This third phase coincided with the introduction of the Open-door policy in the 1970s, and the private investment sector began to participate in land reclamation. Rapid irrigation development influenced the way in which crops were grown on the newly reclaimed land (in particular, new commercial crops were introduced).

Excursus 1

Egypt from the Periphery

One serious problem in the study of the history of modern Egypt is the scarcity of studies based on local documents. Until now, no efforts have been made to research and collect the documents possessed by local notables, because it is believed that Egypt is a centralized country where documents on local societies are collected and held in national archives such as the Egyptian National Archives (Dar al-Watha'iq al-Qawmiya al-Misriya).

This is believed to be the primary reason that little "local history" based on local documents has been recorded. In this context, a series of works by Rudolph Peters (Peters 2007, 2011), who systematically collected local documents in the possession of notables of Qasr village in Dakhla and studied them, are epoch-making.

His works paradoxically show that although—or because—the oasis region is on a periphery far from the centre of Egypt—Cairo—such documents can be found. The authors were also fortunate enough to collect a considerable volume of local documents possessed by village notables during their field surveys of the villages in Dakhla Oasis, especially Rashda, Qalamun, and Balat.

Among the documents collected, there are so-called historical notes or memoirs, which were drafted by local notables, concerning the historical events in the local society of Dakhla Oasis. They are very interesting source materials for the study of Egyptian social history, because the events were described not from the viewpoint of the centre of Cairo but from the periphery of the oasis region.

From this, we have a memorandum comprising 20 documents, in the possession of the present *ma'dhun* of Qalamun. He is from the family of Fuqaha, the family of the *ma'dhun* of Qalamun from *Shaykh* Masud Abdullah in 1913. These 20 documents do not show the name of the writer or the date when they were written; one of them was apparently written by *Shaykh* Masud himself. Its purpose is to record the historical events and accidents that occurred in the central city (*markaz*) and villages in Dakhla Oasis, and it covers the period from 1839 to 1952.

As can easily be imagined, the majority of the memorandum is occupied by the description of the careers of local officials and notables such as their assumption of, and retirement from, official posts, their deaths, the organization or reorganization of local administration, and similar matters. The official posts, which are frequently recorded, are the *hakim* (administrator) and the *qadi* (judge of the religious court) for the regional politics in Dakhla Oasis on the one hand, and the village notables and officials such as the *shaykh al-balad* (village shaykh), the *ma'dhun*, and the umda (village chief).

The *ma'dhun* is the post that is most frequently referred to in the memorandum. In the Nile Valley, the *ma'dhun* was a village notable and was selected according to local customs (*'urf*) to perform civil marriages. The post of the *ma'dhun* has usually been assumed by members of prominent families.

In the oasis region, the New Valley, the *ma'dhun* basically played the same role in Dakhla Oasis. However, he also had the role of the secretary of the religious court (*qalamiya al-mahkama*), drawing up official documents not only for civil marriages but also for civilian affairs such as water distribution, as discussed in Chapter 1.

Another kind of news frequently recorded in the memorandum is of special occasions such as the visit by the king (in 1941) and the ministers (1944) to Dakhla Oasis. The invasion of Dakhla Oasis by Sanusi militants from Libya is a typical case of this kind. In the context of the oasis region of the Western Desert, the Mahdist uprising in Sudan was a serious affair, for their forces captured Baris, a town in Kharga Oasis, in 1893.

However, the memorandum did not record this event, probably because Dakhla Oasis is located inside the Western Desert and was not directly influenced by it. In contrast, one of the main theatres of the Sanusi affair was Dakhla Oasis. In the memorandum, we find the following interesting text. Unfortunately, we could obtain no detailed information on how the Sanusi army occupied Dakhla Oasis, which is contained in a leaflet attached to the end of the folder of the document.

The Sanusi army arrived on January in 1916. In this year, the Sanusi army came from Libya to Dakhla. The men of Mawhub village made it easy for them to occupy the villages of Dakhla. When the Egyptian Army came (in this year), the post of the *ma'dhun* of Mut was assigned to *Shaykh* Sulayman until 1921. During the period from 1916 to 1921, the village of Qalamun had no *ma'dhun;* the post of the *ma'dhun* of Qalamun was given to the *ma'dhun* of Mut, who was *Shaykh* Sulayman.

The information on these events is from so-called official records. However, in the memorandum, in addition to this official information, information on the events and affairs peculiar to the oasis region are included. The most important information of this kind concerns water affairs, some of which are described in Chapter 7. The following is the news of a petition to dig a new well.

On 22 September 1935, the owners of Bir al-Nassajin in Qalamun requested the authority (*markaz*) to dig a well. The request was handed from the umda (village chief) to the police officer (*mu'awin*) and was sent in February 1936. The request was deemed to be an urgent matter in March 1936, because the request had been made repeatedly until 1936.

In addition, we can find information of other kinds, from which we learn about everyday life in the local society, such as unexpected rainfall in 1900, a high rate of mortality probably because of malaria in 1918 and 1922, the first appearance of motor cars in 1925, the introduction of the electric pump for extracting water from a well (in Mut in 1940 and Rashda in 1941), and the rising prices in 1949, after World War II. The following report is of the first appearance of a motor car in the Dakhla region.

At the beginning of the year 1925, the motor car was introduced into Dakhla. The person who brought the first one was *Shaykh* Shadhli Mansur and his distinguished partner (*sharik*), *Hajj Shaykh* Mustafa Jadd al-Rabb, from Jadid. Since then, the motor car has begun to be used for transportation. Before that time, transportation was done with camels and other riding animals.

PHOTO 1
Photo entitled "The view from the house in Rashda village, Dakhla Oasis, where the Grand Sanusi stayed from April to October 1916". According to Hajj Hasan, the Sanusi stayed in Rashda from 1914 to 1916 (Photo taken by Lt. Lindsay; *Source*: Willams 2013: 22).

PHOTO 2
House where the Sanusi stayed

CHAPTER 2

Conditions of Natural and Human Resources

Continuous or sustainable social development can only be achieved when a society's resources are appropriately controlled, used and distributed, since the resources given to a society are limited. The oasis society is particularly so, because of the severe natural circumstances that characterize a desert region. This chapter deals with the natural and human conditions pertaining to resources, specifically water and population, and with the relationship between water, population and land use in the region of Dakhla Oasis, to which Rashda belongs, in order to examine the human activities and the resource conditions under which Rashda lived.

Climate and Hydrogeology

The climate of Egypt is characterized by a hot season from March to September and a cold season from October to February, although the northern winds moderate the temperatures slightly (Robaa, 2008). In the coastal regions along the Red Sea and Mediterranean Sea coasts, average annual temperatures range from a maximum of 37°C to a minimum of 14°C. The most humid area in Egypt is along the Mediterranean coast, where the average annual rainfall is about 200 mm. Rainfall decreases drastically towards the south. It may rain only once in several years in many desert locations (Robaa, 2008). Temperatures vary greatly in the inland desert areas, especially in summer, when the diurnal range is from 7°C at night to 43°C during the day. During winter, temperatures in the desert fluctuate less dramatically, but they can be as low as 0°C at night and as high as 18°C during the day. The average annual temperature increases towards the south from the Delta in the north to the Sudanese border, where temperatures are similar to those of the open deserts to the east and west.

Solar radiation has a magnitude of 12 to 30 MJ/m^2/day, and there are between 3,500 and 4,500 sunshine hours per year (Tadros, 2000). All Egypt is characterized by clear skies during summer (June to August) and partially cloudy skies during the transitional seasons of spring (March to May) and autumn (September to November) (Robaa, 2008). In winter, most of Egypt is characterized by cloudy sky, especially the northern part, while cloud coverage decreases from north to south. A phenomenon of Egypt's climate is the hot spring wind that blows across the country. The winds, known to Europeans as the "Sirocco"

 | DOI 10.1163/9789004317390_004

and to Egyptians as the "Khamasin", usually arrive in April but occasionally occur in March and May. Unobstructed by geographical features, the winds reach high velocities and carry great quantities of sand and dust from the deserts. These sandstorms, often accompanied by winds with speeds of up to 39 m/s, can cause temperatures to rise by as much as 20°C in two hours. The winds blow intermittently and may continue for days, causing illness in people and animals, harming crops and occasionally damaging houses and infrastructure (Robaa, 2008).

Dakhla Oasis is one of the major oases in the Western Desert and is located on low-level ground ranging from 0 to 200 metres above sea level (the surrounding altitude is above 200 metres). The aridity index (AI) used by the United Nations Environment Program (UNEP, 1997) indicates that this area is hyperarid. Here, AI is defined as the ratio of annual rainfall to potential evapotranspiration. Potential evapotranspiration is the amount of water that would be removed from the land surface by evaporation or transpiration if the amount of water already present in the land surface were not a limiting factor. In other words, the potential evapotranspiration over the Sahara desert is very great because the amount of evaporation that could take place there is huge. However, because there is no water there to evaporate, the evapotranspiration that actually takes place is quite small.

In Dakhla Oasis, the annual rainfall is extremely scarce and insignificant (~1 mm/year). Potential evapotranspiration is 1,773 mm to 1,950 mm, calculated using data from the Dakhla meteorological station. This means that the evaporation rate is extremely high in Dakhla Oasis. Thus, the areas around the oasis are bare of vegetation except for very isolated clumps of plants such as *Acacia* or *Tamarix*, which survive on the little moisture that is contained in blown sand and is derived from dew, isolated showers, and frontal storms (Brookes, 2001).

The climate of Dakhla Oasis is mainly controlled by the Harmattan, a dry wind that blows hot air from the Sahara at high pressure. From 1990 to 2005, the relative sunshine duration, which is defined as the percentage of actual sunshine duration to possible sunshine duration, was at a high level of 87% (a low of 82% occurred in January; it peaked at 90% in June) at the Kharga meteorological station 120 km from Dakhla (Robaa, 2008). According to the distribution of annual solar radiation around the world, the highest levels of solar radiation can be found around the southern part of the Western Desert in Egypt (Budyco, 1956).

Although the wind direction frequencies were almost all northerly, there was a slight westerly skewing, probably reflecting a topographic indraft of northern winds into the Dakhla depression (Brookes, 2001). The annual average, maximum, and minimum temperatures were 24.5, 45.3, and −0.3°C respectively in 2008, so the annual range of temperatures is large (the average diurnal range of temperatures was 16.5°C). The annual temperature has increased by 0.038°C per year since 1988.

The annual average relative humidity and vapour pressure were 36% and 10.7 hPa respectively. This means that Dakhla Oasis is humid because irrigation farming has been conducted there for a long time. Since 1988, there has been almost no change in vapour pressure. In terms of human comfort and health, the best seasons are January to March and November to December, calculated using the discomfort index estimated by monthly maximum temperature and average humidity.

The irrigation and cultivation system in the Dakhla Oasis differs from that in the Nile Valley, which depends entirely on the Nile for its water resources. The Nubian Sandstone sequence (Palaeozoic to Lower Cretaceous), which is the world's largest fossil water aquifer system, is its main source of fresh ground-water. The Nubian Sandstone Aquifer System is composed of several laterally and/or vertically interconnected water-bearing strata. It contains fine to coarse-grained continental sandstone from the Lower Cretaceous with good permeability and thickness reaching 3,500 metres (Sefelnasr, 2002, 2007). The average thickness is approximately 1,500 metres (Modified after Heini and Thorweihe, 1993). Its salinity is less than 1,000 ppm.

The recharge areas of the Nubian aquifer are in the north-west part of Sudan and the Uweinat Upland, located at the border between Egypt, Libya, and Sudan. However, the ground-water of the Nubian aquifer is fossil water that was recharged some 14,000 to 30,000 years ago, and the last rainy season related to this recharge occurred 8,000 years ago according to isotopic analysis (Dabous and Osmond, 2001). At present, very little water is recharged by rainfall, so the irrigation water in the Dakhla Oasis is mainly fossil water that has its roots in ancient times.

Demographic Trends

Population trends in Egypt between 1882 and 2006 can be roughly divided into three periods according to the population growth rate. The first period was from 1882 to 1897, when rapid population growth occurred. The population of Egypt in 1882 was estimated to be 5,724,040, which grew to 8,137,417 in 1897 with an annual population growth of 7.3%. However, the 1882 census was an experimental one, and its accuracy is doubtful. Some experts stated that the population had been overestimated.

The second period was from 1907 to 1947. The population grew slowly at 1.2% per year between 1907 and 1937, and 1.7% between 1937 and 1947, from 11,310,399 in 1907 to 12,754,684 in 1917, to 14,425,896 in 1927, to 16,208,112 in 1937, and to 19,138,874 in 1947.

The third period is from 1947 until the present day. The population has grown rapidly since the 1950s. A population that was less than 20 million in 1947 became 25,992,763 in 1960 and reached 30 million in 1966, 47,252,937 in 1987, 59,312,074 in 1996, and 72,798,031 in 2006. The current population is estimated to be 88 million (2015). The population growth rate was 2.4% between 1947 and 1966, and reached a peak of 2.6% between 1976 and 1986. It slowed after 1986, but the growth rate remained high at 2.3% between 1986 and 1996, and 2.1% between 1996 and 2006.

Throughout the 20th century, population growth took place in urban areas. Cairo was the fastest growing city in the 20th century. The population of Cairo governorate was around 50,000 at the turn of the 20th century but reached 6 million in 1986 and 8.5 million in 2006. If the populations of the two governorates of Qalyubiya and Giza, which are part of Greater Cairo, are included, the population of Cairo exceeded 18 million in 2006. The population in 2015 is estimated to be 9,348,501 for Cairo governorate (CAPMAS website).

The population in Cairo governorate grew fastest between 1937 and 1976. Its annual growth reached 4.7% during the period 1937–1947 and was 3.9% in 1960–1966. A natural increase and population inflow also contributed to this rapid demographic expansion.

The population of rural areas, on the other hand, grew slowly until 1976. The rate of population growth stagnated at around 1.0% and only reached 1.5% in 1966. However, this rate was still lower than the national average, implying a high volume of rural to urban migration.

In terms of region, the populations of Lower Egypt and Upper Egypt increased at a higher rate after 1960 but reached a peak in 1986 for Lower Egypt and in 1996 for Upper Egypt.

The demographic trend in the Frontier Governorates, composed of North and South Sinai, Matruh, Red Sea governorates, and the New Valley governorate, is quite different from that in the Nile Valley. The population of the Frontier Governorates has fluctuated drastically over time. For the North and South Sinai governorates, the population increased and decreased suddenly from the 1950s until 1986. This was probably due to the political tension during the Arab–Israeli wars.

As for the New Valley governorate, its population remained stable and did not reach 3 million until 1960. However, it suddenly nearly doubled from 34,000 in 1960 to 60,000 in 1966, making the annual population growth during this period 10.1%.

TABLE 1 *Populations of five oases in the Western Desert (1996)*

	Population			Number	
	Urban	Rural	Total	Cities	Villages
Kharga	49,446	17,641	67,087	Kharga	11
Dakhla	16,252	50,477	66,729	Mut	18
Farafra	2,710	5,248	7,958	Farafra	3
Bahariya	2,114	23,002	25,116	Bawiti	15
Siwa	9,756	3,196	12,952	Siwa	8

Source: Kato and Iwasaki 2008: 17.

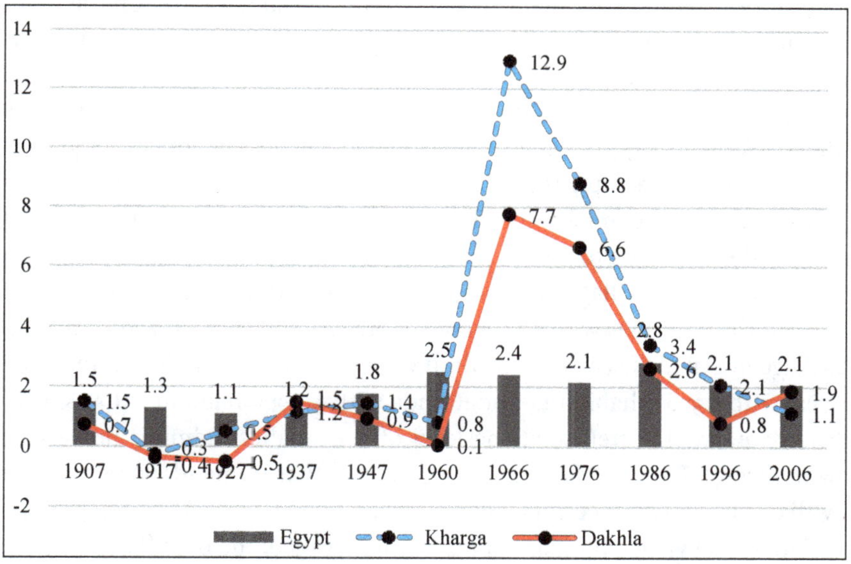

GRAPH 1 *Demographic trends in Kharga and Dakhla Oases and the whole of Egypt (annual population growth %). Source*: Kato et al. 2014: 7.

The New Valley governorate lies in the Western Desert and includes the three oases of Kharga, Dakhla, and Farafra. The Bahariya and Siwa Oases are also located in the Western Desert and are administered by other governorates; Bahariya is in the Giza governorate, and Siwa is in the Matruh governorate. Of these five oases, Kharga and Dakhla are the largest in terms of population size (Table 1).

Graph 1 shows the population trend of Kharga Oasis and Dakhla Oasis from 1897 to 2006. The population growth in the two oases was below the national average until 1960. In fact, the population of the Dakhla Oasis increased only very slowly from 17,090 in 1897 to 21,586 in 1960. Notably, at the beginning of the 20th century, its population growth was almost zero until its recovery in the 1927–1937 period. Its population growth stalled again between 1947 and 1960, making it once again below the national level.

However, Dakhla's population increased suddenly from 1960. The population growth of both the Kharga and Dakhla Oases surged and reached peaks of 12.9% and 7.7% respectively between 1960 and 1966. These rates of population growth were three and six times higher, respectively, than the national average (2.4%) and were obviously due to population inflow from outside the oases. As will be explained below, the principal reason was population inflow enhanced by the New Valley project during the 1960s.[1]

According to the 1996 population census, Kharga Oasis is remarkably urban, with 74% of its population living in Kharga city. In Dakhla Oasis, on the other hand, only 23% of the population live in the urban area (Mut city), and it has more large villages (Table 1).

The populations of oasis villages are small compared with those of villages in the Nile Valley. In 1996, in Lower and Upper Egypt, there were 2,509 and 1,677 villages, with average populations of 7,462 and 8,969, respectively. In these two regions, it is not uncommon for villages to have over 10,000 people. Kharga Oasis, on the other hand, had 11 villages with an average population of 1764. Dakhla Oasis had 18 villages with an average population of 2,969, and Farafra Oasis had three villages with an average population of 2,624. In 2006, the largest village in Dakhla Oasis was Mawhub West with 7,706 inhabitants, followed by Qasr with 5,576 inhabitants, and Rashda with 5,423 inhabitants (CAPMAS).

The demographic trends, at the level of the smallest administrative unit (town/village) (Table 2, Map 3.1, 3.2, 3.3), reveal that tremendous population growth occurred within certain village administrative units. Although population data at the village level are not available for 1960, the data for 1947, 1966, and 1976 suggest that the greatest population growth occurred within certain administrative boundaries.

However, it is noted that the towns and villages that experienced population inflow were not newly created. They existed before the New Valley project was launched. In Kharga Oasis, population inflow occurred mainly in Kharga city and Baris village, between 1947 and 1960, and in Bulaq villages during the period from 1966 to 1976. In Dakhla Oasis, growth was in Mut city, Rashda, Budkhulu, Qasr, Tenida, and Hindawi villages from the 1960s until 1976. After 1976, it was Mut, Qasr, Hindawi, and Balat villages that experienced above

1 On the New Valley project, see Chapter 1.

TABLE 2 *Demographic trends in the administrative units (villages and towns) at Dakhla Oasis (1947–2006)*

	Population					New village created between 1996 and 2006	Population increase (%)		
	1947	1976	1986	1996	2006		1947–1976	1976–2006	1986–2006
Mut	2,513	8,122	11,890	16,252	20,293		223.2	149.9	70.7
Ismant	1,277	2,169	2,737	2,690	2,989		69.9	37.8	9.2
Jadida	2,789	2,991	3,369	3,142	3,662	2	7.2	22.4	8.7
Rashda	1,739	4,398	5,574	4,364	5,423		152.9	23.3	-2.7
Qalamun	1,587	2,272	2,184	2,128	1,812		43.2	-20.2	-17.0
Sheykh Wali		1,721	2,054	1,990	2,376			38.1	15.7
Awina				1,659	1,884				
Qasr	2,155	5,813	5,257	3,723	10,122	9	169.7	74.1	92.5
Maasara	1,202	2,374	3,031	3,282	3,911	2	97.5	64.7	29.0
Mushiya	1,146	1,571	1,778	2,146	2,495		37.1	58.8	40.3
Hindaw	1,038	2,174	2,774	3,062	3,942		109.4	81.3	42.1
Budkhulu	537	1,113	1,216	1,523	1,719		107.3	54.4	41.4
Balat	2,668	4,952	5,492	5,126	6,397	8	85.6	29.2	16.5
Tenida	1,020	2,261	2,774	3,114	3,403		121.7	50.5	22.7
Izab Qasr	1,711	3,210	3,872				87.6		
Mawhub				1,871	1,678	2			
Mawhub West			3,879	6,278	7,706				98.7
Total	21,382	45,141	57,881	62,350	79,812		111.1	76.8	37.9

Source: Kato et al. 2014: 7.

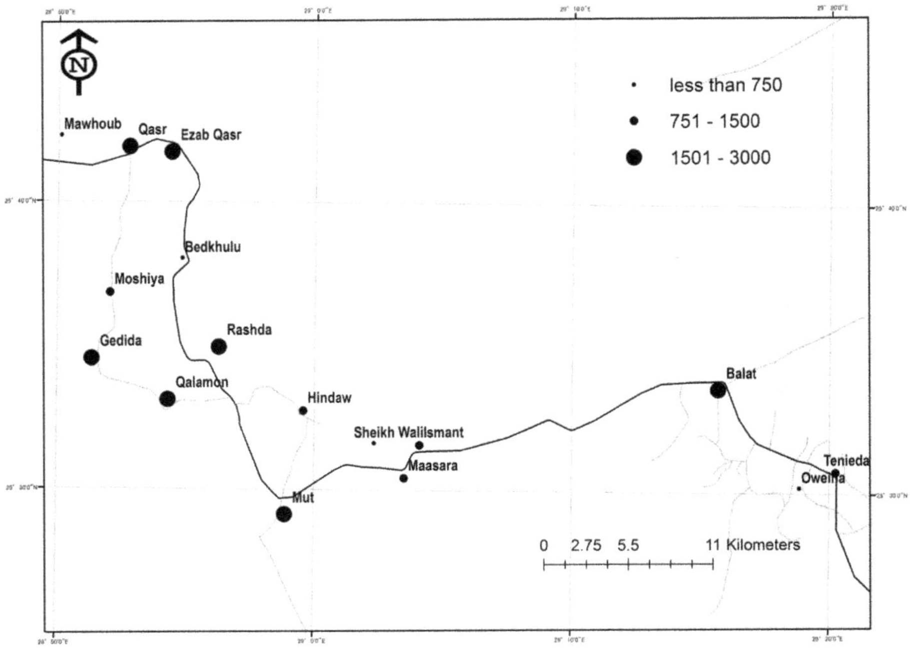

MAP 3.1 *Population by administrative unit (1947). Source*: Kato et al. 2014: 7.

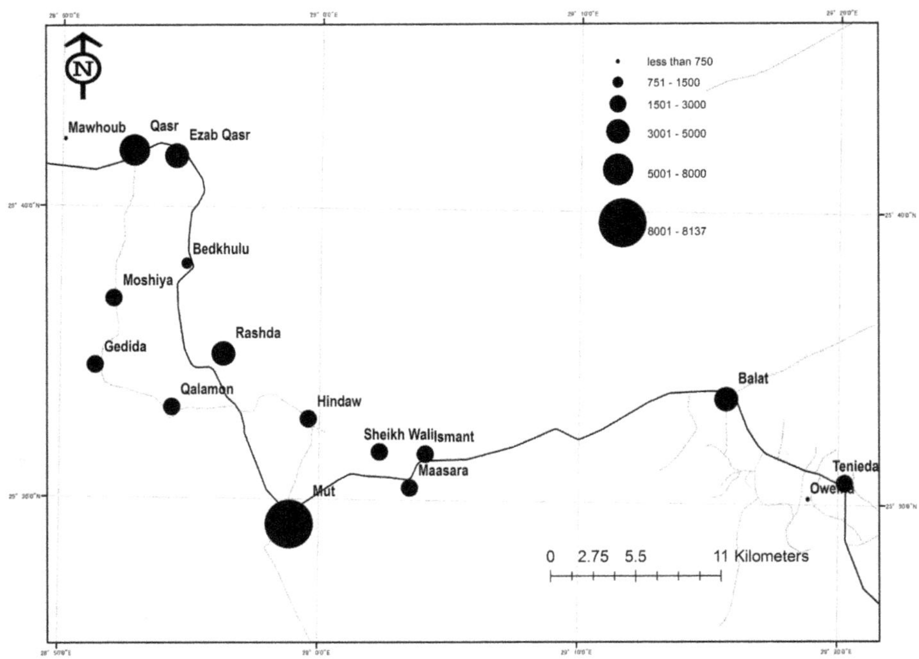

MAP 3.2 *Population by administrative unit (1976). Source*: Kato et al. 2014: 7.

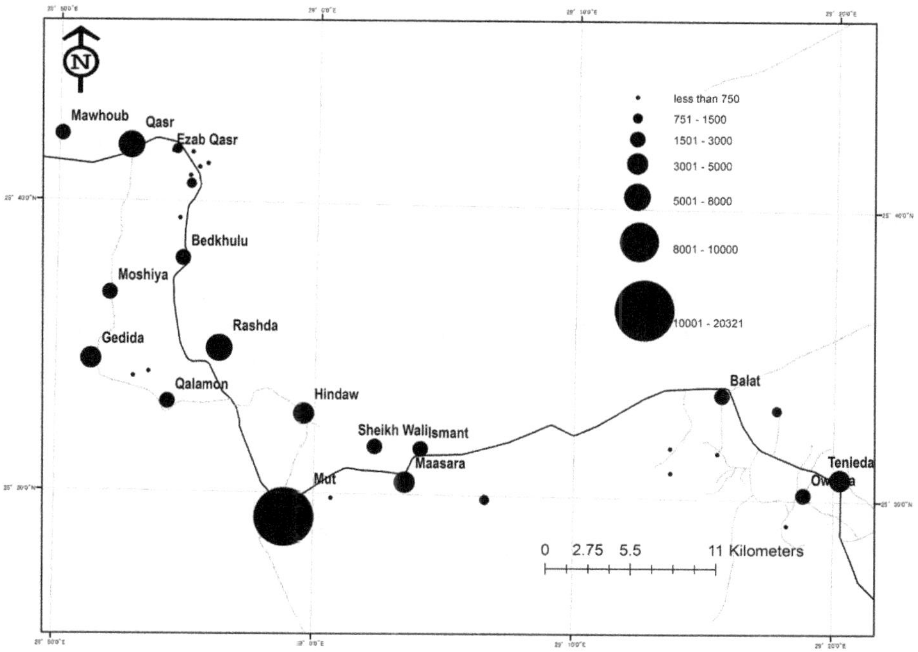

MAP 3.3 *Population by administrative unit* (2006). *Source*: Kato et al. 2014: 7.

average population growth, especially between 1996 and 2006. In Balat and Qasr villages, many new administrative units were created between 1996 and 2006, but they became administratively independent from these two villages.

The sharpest increase took place in the administrative centres of both oases. This was in Kharga city in Kharga Oasis and in Mut city in Dakhla Oasis. Kharga city was a small village with a population of less than 5,000 until the 1950s. However, its population now exceeds 50,000. Mut was a small village with fewer than 3,000 inhabitants until the 1950s. Now its population is more than 20,000. Therefore, the demographic explosion in Dakhla Oasis occurred within the pre-existing town and village boundaries, mainly because of the migration of people attracted by the development of administrative centres.[2]

2 Age Structure. In terms of age structure, the inhabitants of oasis villages have a higher mean age than those of villages in Lower and Upper Egypt. The villages in Lower and Upper Egypt have a larger proportion of the population under 15, whereas those in Kharga and Dakhla Oases have a larger number of inhabitants aged between 25 and 34, and 65 or above. The villages in Farafra are different from those in the other two oases. They have few inhabitants above the age of 55.

Water, Population, and Land Use

As mentioned above, the ground-water of the Nubian aquifer, which is the water source for Dakhla Oasis, is fossil water. This exploitable ground-water resource is very large. However, because there is essentially no rainfall in this region and the water is not rechargeable, the wells are destined to dry up in time.

However, the ground-water or aquifer level differs depending on the geological conditions underground, so although ancient wells and the land around them are continually abandoned, new wells continue to be exploited. The pressure in the aquifer has been greatly reduced; most of the springs (*'ain*) have become dry, and some wells are no longer flowing, but other wells are still flowing (Diab 1984). Thus, the population has been concentrated close to the ground-water wells for many decades, and when the wells become dry, the people move to another area looking for water supplies.

At the beginning of the 20th century there were 420 ancient wells in the Dakhla Oasis, known by the natives as *'ain rumani*, and 162 modern wells, called *bir*, or *abyar* in the plural.[3] Now, most of these natural springs and free-flowing shallow wells that used to feed the oasis have dried up. Instead, there are 523 private and/or shallow wells and 313 deep wells (700–1,200 metres) extracting the upper and lower aquifers in the Dakhla Oasis (Sefelnasr 2007: 100).[4]

However, recently, because of the increased population and agricultural expansion, the amount of water discharged has increased at a higher rate than the aquifer has been recharged. The ground-water of the Nubian Sandstone aquifer has been heavily exploited since 1960. The amount of water discharged has increased tremendously as shown by the extraction rate of water (cubic metres per day) between 1960 and 1997 (Graph 2).

The impact of this heavy exploitation is clear in the continuous lowering of the ground-water level in wells within the oasis, increased salinity, and changes in the hydraulic behaviour of some wells. Some artesian flows have ceased, especially on the eastern side of the oasis. Dozens of shallow wells have already dried up. The aquifer is still being stressed by pumping from wells that double in number from one decade to the next. The development ambitions of both

---

3 In 1874, Rohlfs reported that a man named Hasan Effendi, originally working with the French mining engineer LeFèvre, had drilled 60 wells in Dakhla in 30 years. See King 1917: 358.

4 According to Gad et al. (2011), there are about 13 productive deep wells discharging approximately 30,703 m^3/day, and the total cultivated area is about 565 hectares. Himida (1966) expected that the deep wells in Dakhla would stop flowing by 1987.

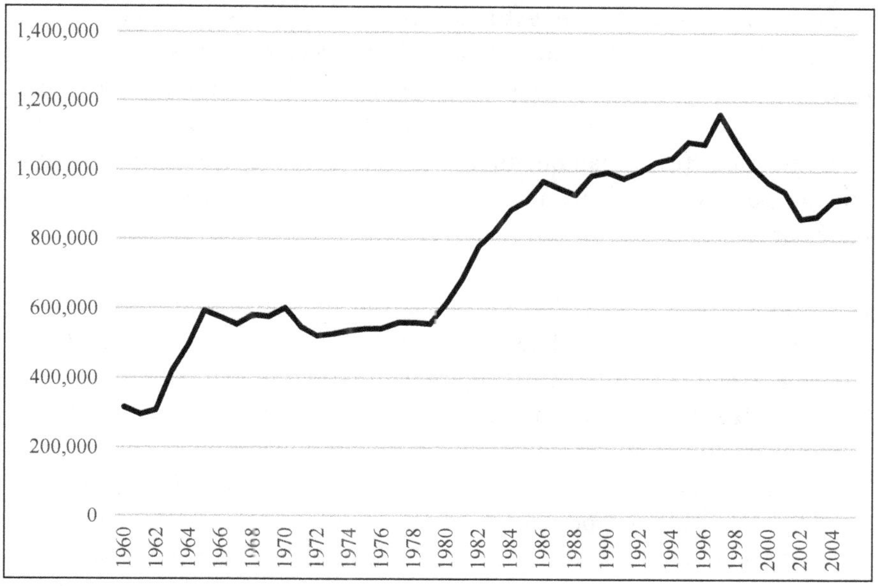

GRAPH 2 *Extraction rate in Dakhla Oasis (1960–2005) (cubic metres per day). Source*: Based on Sefelnasr 2007: 159–161.

the government and the residents have led to increasing utilization of the ground-water resources.

Unmanaged ground-water extraction in Dakhla Oasis has led to several changes that can be seen in the hydraulic head patterns and flow directions: local flow patterns, and noticeable cones of depression that have appeared in the past 50 years (since the beginning of the New Valley project in the 1960s).

Continued uncontrolled pumping will lead to shallow wells becoming completely dry and increases in the depth required to reach ground-water. In the case of deep wells, when the water level declines, the well becomes inoperable because of the position of the water pump in the aerated zone (above the dynamic water level). Therefore, the owner (government or investor) drills a new well (replacement well) in the same area or nearby to replace the original one.[5] However, continued pumping will become too great for extraction to be economical.

5 Given the soil suitability, topographic features, and recent hydrological conditions in the replaced well, a new design was developed to adjust the position of the screen intervals and the depth at which the submerged water pump was installed.

Unmanaged ground-water extraction has also caused social and economic changes. The ongoing dynamic relationship between ground-water resources, land use/land cover, and demographic trends in Dakhla Oasis since the 1970s can be analysed by integrating remote sensing and through geographic information system (GIS) manipulation. The data that are used in the analysis are the population census and the following three sources of geographic information.

1. Satellite data: The satellite data used were one Landsat Multispectral Scanner System (MSS) image acquired in Nov. 1972, two Landsat Thematic Mapper (TM) images acquired in November and December 1984, and six SPOT4 satellite images acquired in August and September 2011.
2. Site investigation data: A total of 217 ground truth points for different classes of land use and land cover were collected during fieldwork in April 2013. These ground truth points are distributed among all the villages of Dakhla Oasis. The assigned land use classes were urban, bare land, perennial crops, annual crops, bodies of water, reclaimed land, and *sabkha* (salt flats).
3. Topographic and hydrogeological maps: Four 1:100,000 topographic maps centred on Rashda, Mut, Balat, and Qasr villages, respectively, issued in 1986, one 1:25,000 historical topographic map of Rashda issued in 1932, and one 1:500,000 hydrogeological map of the Dakhla Oasis issued in 1998 were used, in addition to field data and images, to support the mapping of land cover changes in Dakhla Oasis.

The false colour composites generated from bands 4, 3, and 2 of the three images were visually interpreted through on-screen digitizing. The visual interpretation provided a general idea about land cover changes over the study period. Many urban settlements were erected in the oasis during the period.

A noticeable change is detected in areas of land reclamation in the study area, where part of the desert was converted to cultivated lands. On-screen digitizing was conducted for urban, *sabkha*, and salt marsh land cover classes as well as the road network and was integrated with the results of classification.

Supervised classification using all reflective bands of the three images, acquired in 1972, 1984, and 2011, was conducted using a maximum likelihood classifier. To increase the accuracy of land cover mapping of the three images, ancillary data and the results of on-screen interpretation were integrated with the supervised classification results.

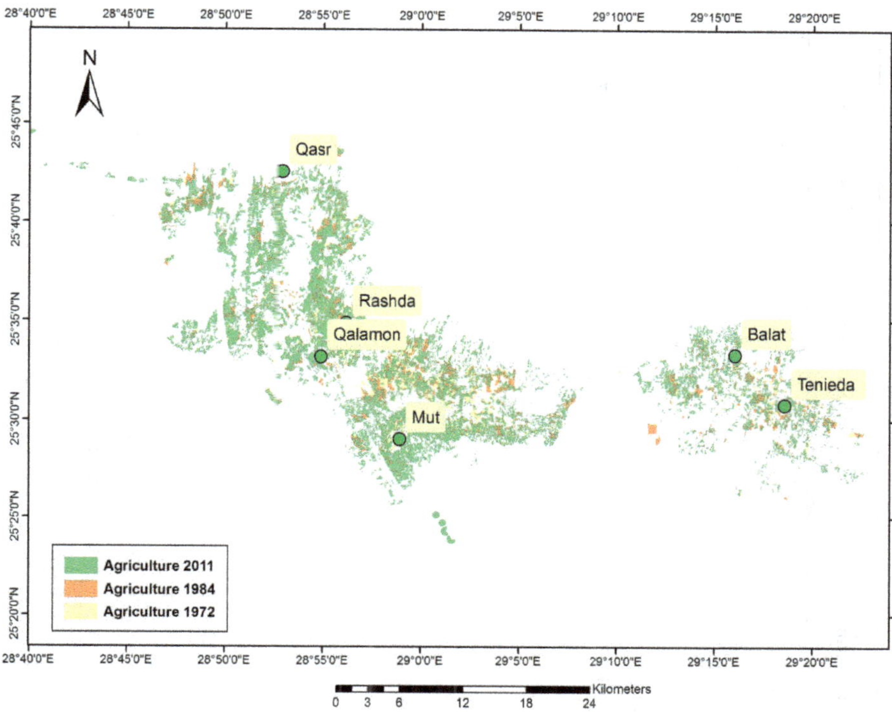

MAP 4 *Change in distribution of agricultural areas in Dakhla Oasis villages from 1972 to 2011. Source*: Kato et al. 2014: 8.

This overlay of the visual interpretation on the results of classification led to increases in overall accuracy of approximately 10% for the three images. The benefits of adding the on-screen interpretation are attributable to spectral confusion of the urban settlements with the bare land class, because the buildings are made from the surrounding local materials.

Post-classification change detection was conducted using a cross-tabulation GIS module on the classification results of the 1972, 1984, and 2011 images, to produce change images (Map 4) and statistical data (Table 3).

Table 3 compares the rates of land cover change with the rates of population change for Mut, Balat–Tenida, and Qalamun–Rashda for the periods 1984–2011 and 1986–2006. For example, the total agricultural land cover increased from 109.52 square km in 1984 to 209.29 square km in 2011. Total agricultural land cover increased by about 90% between 1984 and 2011. Total urban land cover increased from 5.28 square km in 1984 to 11.38 square km in 2011, a growth rate of about 115% during this period. In addition, the figure shows that the total population increased in the period 1986–2006 from 27,914 to 37,328 with a

TABLE 3 *Change in extent of agricultural land and urban settlements in Dakhla Oasis villages between 1984 and 2011, and changes in population between 1986 and 2006*

	Agricultural area (km^2)		Urban area (km^2)		Population		% increase between 1984 and 2011		
	1984	2011	1984	2011	1986	2006	Agric	Urban	Pop
Balat & Tenida	16,983	36,734	0,427	1,527	8,266	9800	116,29	257,61	18,55
Mut	36,846	69,254	3,486	6,679	11,890	20293	87,95	91,59	70,67
Qalamun & Rashda	55,694	103,31	1,369	3,175	7,758	7235	85,49	131,92	-6,74
Total	109,523	209,298	5,282	11,381	27,914	37328	91,10	115,46	33,72

Source: Kato et al. 2014: 7.

TABLE 4 *Total Extraction in cubic metres per day in Balat–Tenida, Mut, and Qalamun–Rashda Villages*

	Extraction (m^3/day)			% change in extraction		Population			Population increase (%)	
	1976	1986	2006	(1976–2006)	(1986–2006)	1976	1986	2006	(1976–2006)	(1986–2006)
Balat & Tenida	114,508	204,786	178,360	56	-13	7,213	8,266	9,800	36	19
Mut	68,487	122,481	105,834	55	-14	8,122	11,890	20,293	150	71
Qalamun & Rashda	42,982	71,504	69,915	63	-2	6,670	7,758	7,235	8	-6,74
Total	225,978	398,771	354,109			22,005	27,914	37,328		

Source: Kato et al. 2014: 8.

growth rate of about 34%, which accompanied an even greater increase in urban land cover and agricultural land.

From Table 4, it is clear that the overall discharge increased from 1976 to 2006 by 55% to 63% (depending on area), which is associated with an increase in population ranging from 8% to 150% in the same period. In contrast, from 1986 to 2006, the discharge decreased by between 2% and 14%, while the population continued to increase by 19% to 71%. This decrease in discharge in the past 20 years could be attributed to some combination of the following factors: 1. the use of modern methods to irrigate cultivated lands including sprinklers and drip irrigation, 2. lowering of the ground-water table, and 3. drying up of some wells.

CHAPTER 3

Family Structure

Family is a key to understanding Arab and Egyptian societies. Fortunately, we had the opportunity of conducting surveys in 19 villages, as mentioned in the Introduction. Through these surveys, micro data at the household level were collected. In addition to the household surveys, we made maps of residential areas for each village, so as to link the household data to the geographical data. In this chapter, we introduce our analytical framework and then using these micro data and information, examine the structures of household and family in the villages of Dakhla Oasis, to which Rashda belongs, and compare them with villages in other regions of Egypt.

Analytical Framework

Basic Concepts

Family is a favoured topic in socio-economic studies of Arab or Middle Eastern societies. Studies of Egypt are no exception[1]. This is because families are assumed to have an important role in social life and have been viewed as a prototype of the societies concerned.

Many studies of rural Egypt point out the importance of the family in social life. To investigate this issue, we first distinguish household and family. A household by definition is a residential unit where individuals share consumption. It is usually called *usra* in Arabic, although it is called *bait*, *manzil*, and other terms in documents such as population censuses in old times. *Bait* and *manzil* literally signify "house".

We address these two concepts of household and family, with the assumption that the basic unit of material life is the household. Because families in social life are based on households, the analysis in our study must be based on households. In fact, the household is the most commonly used unit in statistics on the national scale. Therefore, for practical reasons, we must use the household as the unit of analysis to link our village-level data with national statistics.

1 A comprehensive description of the family in modern Egypt was published in 2015 by Kenneth CuNo. Our special concern is with the family structure in villages, which is described in Chapter 2.

 | DOI 10.1163/9789004317390_005

On the other hand, a family is a social unit based on a household. In Egypt, the family is generally called aila (*'a'ila'*) in Arabic. Our primary interest is in understanding the social life of families in a way that goes beyond their material life. The household surveys in 19 villages mentioned in Introduction were conducted for this purpose. To study families, the concept of family must be defined, which is a difficult task because it is a flexible concept. Therefore, we preferred to define the family according to what the households understand, by asking the question "what is your *laqab al-'a'ila*?". In this sense, *laqab* means the family name, as opposed to *ism*, which is a given or first name.

Review on Family Studies

Family is a somewhat vague concept, because the role of family is defined differently according to the culture of each society. First, a family is a kinship group. In fact, studies of rural societies in North Africa, including Egypt, point out the importance of kinship or family (aila), translated as the extended family or patrilineal group.

In political studies, kinship has been considered to be a major force in shaping the village politic (Antoun and Harik 1972, Gellner and Waterbury 1977). In sociological studies, it has been considered to be a benchmark of rural transformation. The dissolution of aila has been described as the cause or result of social changes in villages, which have led to the phenomenon of modernization of rural societies.

However, despite their acknowledgement of the importance of kinship, few studies have analysed its socio-economic function. Kinship has been viewed, explicitly or not, from a solely normative perspective; it is seen as a basic social unit that determines the behaviour of members who supposedly share the collective values to which their kinship group adheres.

This view of kinship was particularly strong in the literature of the 1960s and 1970s. These studies, influenced by literature from the colonial period such as J. Berque's study of Moroccan and Egyptian villages, have regarded the kinship group of aila as the basic unit of village structure (Berque 1955, 1957). Such groups have been assumed to be composed of a patriarch, his spouse, and his unmarried and married descendants, forming a community located in different quarters of the village but strongly bound by a collective norm. Thus, kinship has been considered to be important in defining patriarchality, patrilineality, and patrilocality.

Not only anthropological studies, with a tradition of studying lineage, but also the socio-economic studies during this period have focused on kinship (Duvignaud 1968, Baer 1969, 1976). French and Japanese scholars have been the most eager to do so. They have attempted to study village structures through the analysis of kinship relations in land management, exploitation, and non-agricultural activities (Chaulet 1987, Nakaoka 1973, Kimura 1973). Their studies assumed that social and economic relations in the villages are a matter of power relations between the different aila.

Thus, the Japanese studies focusing mainly on Egypt, as well as French studies focusing on the Maghreb (Tunisia, Algeria, Morocco), have acknowledged the importance of aila in determining the socio-economic behaviour of villagers and have set aila as a unit of analysis. This can be called a "family exploitation" or "family strategy" approach or one of *kyodoutai-ron* ("theory of community"), using the French and Japanese terms, respectively.

Criticism of the Kinship Approach

Today, this approach is criticized on two grounds. First, the notion of the kinship group as opposed to the individual or nuclear family is claimed to be a fictive one, created by colonialists to bestow legitimacy on discourse about the backwardness of indigenous societies (Charrad 1990, 2001). Second, the kinship organization is instead a result of the social and economic interactions among individuals who use the kinship relations as a strategy.[2]

In both anthropological and sociological studies, scholars have argued for the cultural re-interpretation of kinship patterns (Geertz et al. 1979) and the need to build theoretical bridges between individuals/households and kinship to attempt a conceptualization of kinship relations in terms of "relativity" or "family networks".[3]

On the other hand, in socio-economic studies of rural societies in North Africa, especially Egypt, such attempts have been difficult, mainly because of the lack of data. Micro data at the household level, especially in regard to land and cultivation, have been limited. In addition, until the beginning of economic liberalization in the 1980s, the state had been conceived as the main driving force of rural development.

Such a view, together with the scarcity of data, led scholars interested in rural societies to focus on agricultural policy or state intervention (Saad 1988,

2 Bourdieu's early study, for example, has criticized his own study of kinship relations in the Algerian countryside and argued that, in reality, the symbolic lineage relations do not determine kinship relations (Bourdieu 1980). He argued that villagers who negotiate with each other based on economic and political motivations determine the actual kinship relations.

3 Such a trend can be seen in works by authors such as Hopkins (2003) and Ben Salem (1994) on family sociology in Arab societies in general, Miyaji (1993) on migration in rural Algeria, and Joseph (1999), who studied identity, as well as Singerman and Hoodfar (1996) and Singerman (1997), who dealt with gender in urban societies, especially Cairo. For details, see Iwasaki (2006).

Dyer 1997). From the 1980s on, with the withdrawal of the state, several scholars have focused on "farmers" instead of the state. "Farmer" is contrasted with "peasant", which is a concept more closely linked to traditional relations of kinship and land (Mehenna et al. 1994, Hopkins 1993).

However, these studies do not take kinship into account in their analyses. They claim, instead, that villager behaviour can be explained exclusively by economic rationality. Thus, in socio-economic studies, the challenge to connect the kinship pattern to the behaviour of individuals and households still remains unmet.

Family as Social Institution

An overview of the literature can be summarized as follows. First, there is a lack of socio-economic studies on family in rural societies. This is primarily because of data constraints and the economic context in which the agricultural policy and state intervention attracted scholars' interest during the period of the planned economy. In contrast, in the era of marketization, the "farmers'" economic behaviour attracted the scholars' attention as a driving force of rural development.

Second is the analytical framework that sets the family as a unit of analysis. This framework is not appropriate for studying the dynamics of rural transformation, because it disregards the peasants' economic behaviour, which shapes family patterns. The family should be analysed as a social institution in which the individual or household acts strategically.

The institutional approach would be efficient for such an analysis of families. It is an approach that can analyse the family as an outcome of individual actions in the nexus of socio-economic interests, in contrast to the approach taken in the study of family exploitation or strategy in France and *kyoudoutairon* in Japan. The latter approach explains individual behaviour in terms of collective value, whereas the first approach considers how individual behaviours form an institution.[4]

Referring to a system of joint exploitation between family (aila) members that existed in 19th-century Egypt, Kato (1983) criticized the prevailing approach among Japanese scholars focusing on aila to provide the logic for explaining the socio-economic phenomenon. He argued that this customary pattern of joint exploitation did not appear solely because of the state's intention to limit land-holding rights to the head of an aila, so as to facilitate taxation. He argued that such arrangements were partnerships contracted

4 For our purposes, one of the best studies of Egyptian society from an institutional approach to family is Binder (1978).

between peasants to share the land, labour, and capital for joint management, which functioned as a form of profit sharing and risk management.

Two conditions are required in order to apply the institution approach. First, the household, defined by co-residence and the sharing of a budget, should be set as the unit of analysis, instead of family or kinship. Setting the household as the basic unit of decision-making in production and consumption has the advantage of analysing the family as a social institution. If the household were the basic unit of analysis, the family would be analysed as a product of arrangements between actors with different rights and interests, rather than as natural or socially given.

Second, family relations should be considered to be a resource mobilized by individuals/households. Such a resource could be called "social capital". This enables possible analysis of what kinds of individuals/households utilize the family network, and in what circumstances, which would be indicators of the socio-economic functions of the family.

Household Structures in Nationwide Surveys

Our aim is to study the socio-economic functions of the family. In this regard, taking account of regional diversity is important. Every region has its own socio-economic setting; therefore, we presume that the functions of a family differ accordingly.

Of course, everybody interested in Egypt knows that it is regionally diverse. It is well known that there is a wide disparity between urban and rural Egypt, and that Cairo, the capital city, is outstanding in terms of social and economic life. It is known also that rural Egypt, and Lower and Upper Egypt differ economically, socially, and culturally. In fact, the following two nationwide surveys, which provide the only information available on families, illustrate the differences within rural Egypt.

The Demographic Health Survey (DHS) is the most detailed household survey in Egypt. It is conducted nationwide every three to five years, and it provides detailed information on fertility, consanguinity, and other demographic features of households. Many demographers and sociologists used the DHS to understand the households of Egypt, mainly with respect to their demographic aspects.

According to the figures of the latest DHS in 2014 concerning prevailing fertility levels, rural women will give birth 3.8 times by the end of their childbearing period, while for urban women, the figure is 2.9 births.[5] By place of residence, the current fertility level varies from 2.5 births in the Urban Governorates to 4.1 births in rural Upper Egypt (EDHS 2014: 9).

Marriages between relatives (consanguineous marriages) are common in Egypt. According to the 2008 Demographic Health Survey, 34.4% of currently married women reported that their current husband, or in the case of widowed or divorced women their most recent husband, was a relative. Most consanguineous marriages involved first or second cousins. In such marriages, the husband was somewhat more likely to be a relative from the father's side than the mother's side (14% and 8%, respectively) (Zanaty and Way 2008: 98).

Consanguineous marriages were more common among rural than urban women; one-third of the marriages in rural areas involved relatives, compared with fewer than one-quarter of marriages in urban areas. Considering place of residence, the highest rate of consanguineous marriages was found in rural Upper Egypt, where nearly one-half of marriages were between relatives (Zanaty and Way 2008: 99).

Another piece of information is from the population census sample data provided by the Minnesota Population Center's Integrated Public Use Microdata Series, International (IPUMS-I). These are raw data, 10% of which are extracted from the census, and they contain some variables related to household structure, fertility, or marriage. Table 5 shows the classification of households into nine types. Households in Egypt are obviously of a nuclear family type, because households composed of one couple, or one couple and their unmarried children, represent 77.2% of Egyptian society.

However, there is a considerable difference as to the proportion of the "extended family" type. Here, "extended family" is a household with member(s) who are relatives other than spouses or children. Couples living with their parents, brothers, or sisters fall into this category. This extended family type occupies 11% of all households in Egypt. There is a substantial difference in its proportion across and within the regions.

5 According to the latest Demographic Health Survey in 2014, reversing the downward trend that was generally observed throughout the period, the TFR increased between the 2008 and 2014 Egypt Demographic Health Surveys by 17 per cent, from 3.0 births to 3.5 births. The 2014 TFR is at the same level as the TFR observed 14 years ago in the 2000 EDHS (EDHS 2014: 7).

TABLE 5 *Household types in Egypt by governorate (%) 2006*

Region	Governorate	One-person household	Married/ cohab couple, no children	Married/ cohab couple with children	Single-parent family	Polygamous family	Extended family, relatives only	Composite household, family and non-relatives	Non-family household	Group quarters	Unclassified subfamily	Total	Number
Urban Governorates	Cairo	2.6	5.3	70.1	8.0	0.1	10.9	0.6	0.4	1.0	1.0	100.0	790,356
	Alexandria	1.9	5.9	74.4	7.2	0.3	8.4	0.3	0.2	0.6	0.8	100.0	413,185
	Port Said	1.6	4.8	66.1	7.1	0.2	15.0	0.7	0.5	1.9	2.1	100.0	56,963
	Suez	1.6	4.0	69.6	8.2	0.2	12.5	0.7	0.4	0.8	2.1	100.0	51,282
Lower Egypt	Damietta	1.9	5.8	76.8	7.5	0.3	6.1	0.2	0.2	0.3	0.8	100.0	109,528
	Daqahliya	1.7	5.2	72.8	8.5	0.2	9.4	0.3	0.1	0.3	1.6	100.0	499,164
	Sharqiya	1.4	4.4	74.4	6.8	0.3	10.6	0.3	0.1	0.2	1.4	100.0	535,778
	Qalyubiya	1.5	5.0	77.5	6.3	0.3	8.1	0.4	0.1	0.2	0.7	100.0	424,180
	Kafr Sheykh	1.1	4.2	72.0	6.6	0.8	14.1	0.2	0.1	0.3	0.7	100.0	262,222
	Gharbiya	1.6	4.6	73.4	7.7	0.3	11.3	0.3	0.0	0.3	0.6	100.0	401,449
	Menufiya	1.3	4.3	74.5	6.3	0.4	12.0	0.3	0.1	0.3	0.6	100.0	327,229
	Buheyra	0.9	4.3	72.9	5.3	1.0	14.0	0.3	0.0	0.1	1.2	100.0	474,888
	Ismailiya	1.5	4.6	72.2	7.0	0.5	12.3	0.5	0.2	0.4	0.8	100.0	95,327
Upper Egypt	Giza	1.6	5.2	77.2	6.2	0.4	7.4	0.5	0.3	0.6	0.7	100.0	629,420
	Beni Suef	1.3	3.6	69.2	7.6	0.6	16.8	0.1	0.1	0.2	0.5	100.0	229,307
	Fayyum	1.5	4.1	76.4	7.6	0.6	9.2	0.1	0.0	0.2	0.4	100.0	251,518
	Minya	1.4	3.9	71.3	6.9	0.6	14.0	0.3	0.1	0.4	1.2	100.0	416,558

	Asyut	1.3	4.0	72.4	8.8	0.5	10.1	0.3	0.2	0.8	1.6	100.0	345,729
	Suhag	1.5	4.4	69.7	11.9	0.9	9.9	0.2	0.0	0.3	1.3	100.0	375,022
	Qena	1.3	3.6	65.2	10.0	0.8	16.5	0.3	0.3	0.3	1.7	100.0	299,610
	Aswan	1.7	4.1	68.9	10.1	0.5	11.7	0.3	0.3	0.9	1.5	100.0	118,859
	Luxor	1.4	4.2	70.6	7.0	0.6	14.0	0.0	0.1	1.7	0.5	100.0	45,367
Frontier Governorates	Red Sea	2.5	4.4	59.2	4.1	0.4	5.9	0.3	3.9	19.1	0.3	100.0	28,886
	New Valley	1.1	4.7	69.3	5.3	0.6	16.4	0.1	0.2	2.0	0.3	100.0	18,655
	Matruh	0.5	3.1	66.7	3.8	3.3	18.7	0.7	0.7	1.1	1.5	100.0	32,316
	North Sinai	1.0	4.6	71.2	7.0	1.2	11.1	0.8	0.7	0.9	1.7	100.0	34,569
	South Sinai	2.3	3.2	36.5	2.9	0.3	4.8	0.6	6.8	42.3	0.5	100.0	15,067
Total		1.6	4.6	72.6	7.5	0.5	11.0	0.3	0.2	0.6	1.0	100.0	7,282,434

Note: A "family" is any group of people related by blood, adoption, or marriage. An unrelated individual within the household is considered to be a separate family. Thus, a househ consisting of a widow and her servant contains two families; a household consisting of a large, multiple-generation extended family with no lodgers or servants would count a single family (IPUMS-I).

Source: CAPMAS. Population, Housing and Establishment Census 2006 sample data. Minnesota Population Center Integrated Public Use Microdata Series, International (IPUMS Version 6.3 (Machine-readable database). Minneapolis: University of Minnesota, 2014.

Household Structures and Family in 19 Villages

Type of Household

The next step was to examine data from 19 survey villages in Egypt; first, those on household structures; and second, those on family structures. In the 19 villages, the most common type of household is the nuclear family household (parents and children) (Table 6). On average, 64% of the households in the 19 villages are of the nuclear family type; 9% are couples without children.

Other household types also exist, although their numbers are few (27.9% on average). These types are single-person households (4%) or a household with one parent and children/siblings, or a nuclear family living with at least one parent or other relatives.

Some villages have a higher proportion of other household types. One type is Awlad Sheykh, which has 30.2% of households, and Sheykh Isa in Qena governorate (Upper Egypt), which has 20% of households composed of a female household head and children. As mentioned below, this may be related to the fact that Awlad Sheykh has many migrants abroad.

Another type of household is the extended family type, which accounts for more than 20% of households in Hamra in Kafr Sheykh governorate (Lower Egypt) and Beni Samrag in Minya governorate (Middle Egypt). In these two villages, many households have a married son living with his parent(s), or married brothers and their wives living together.

To understand the details of household structure, we calculated the number of relatives per 100 households. The results shown in Table 7 reveal two essential relationships that cause variations in the living arrangements. One is the vertical relationship, which is that between a parent (father/mother) and son. Another is the horizontal relationship, which is that between two brothers (or a sister). In all villages, the living arrangements follow parent–son relationships and brother–brother/sister relationships. This fact illustrates the importance of patrilineality as a principle of household organization in rural Egypt.

Importance of the vertical and horizontal relationships is higher in some villages. People in the three villages of Hamra, Beni Samrag, and Sheykh Isa are more likely to live with parents (father/mother) and brother/sister. For instance, 14% of household heads in Hamra live with their child's spouse, 26% with their father or mother, and 27% with brother or sister. Thus, the households in Hamra tend to be larger, with an average of 6.3 people per household.

In contrast, households in the two villages of Zahra and Abu Tawala in Sharqiya governorate (Lower Egypt), Kafr Beni Hilal in Buheyra governorate

TABLE 6 *Household composition in 19 villages (2005)*

	Household type							
	Parents + Unmarried children	Couple	Other	Total	Number of households	Household size	% of female household head	Average age of household head
Abu Senita	60.4	10.8	28.8	100.0	600	4.8	11.3	47.8
Zahra	72.5	11.8	15.7	100.0	600	4.1	10.7	44.3
Abu Tawala	66.3	13.2	20.5	100.0	600	3.9	16.8	47.7
Kafr Shubrahur	66.7	8.4	24.9	100.0	142	4.3	16.9	47.4
Hamra	51.4	2.9	45.7	100.0	1,000	6.3	13.3	48.5
Kafr Beni Hilal	70.6	8.1	21.2	100.0	595	4.2	15.5	44.4
Sidi Oqba	64.9	7.8	27.4	100.0	600	4.8	13.7	47.5
Homa	68.3	8.1	23.6	100.0	600	5.3	10.8	43.8
Damashqin	68.0	4.0	28.0	100.0	600	5.5	12.0	46.6
Beni Samrag	52.3	6.6	41.1	100.0	600	6.2	16.2	47.8
Kawamel Bahry	69.1	12.9	18.1	100.0	600	4.8	7.2	45.0
Awlad Sheykh	43.4	10.6	46.1	100.0	600	4.6	37.5	41.4
Sheykh Isa	47.9	6.3	45.8	100.0	600	5.6	20.2	53.4
Munira	72.6	7.7	19.7	100.0	300	4.4	7.3	50.7
Bulaq	70.0	8.1	21.9	100.0	301	4.4	9.6	50.5
Rashda	74.7	12.4	12.9	100.0	550	4.2	4.6	46.8
Balat	74.6	10.8	14.5	100.0	550	4.1	8.4	48.1
Total	63.6	8.6	27.9	100.0	7,638	4.9	13.9	46.8

Source: 2005 Rural Household Survey data.

TABLE 7 *Number of relatives per 100 households by age of the household head (unit: person) (2005)*

	15–24	25–34	35–44	45–54	55–64	65 & above	Total
Child's spouse	0	0	0	5	18	25	8
Father/mother	21	11	9	7	3	0	7
Brother/sister	41	13	8	4	2	3	7
Grandchild	0	0	0	6	29	61	15
Other relative	1	2	3	3	2	1	2
Other	0	0	0	0	0	0	0
Total	3	4	5	6	5	4	5
Number of households	269	1,743	2,546	2,256	1,596	1,328	9,738

Source: 2005 Rural Household Survey data.

(Lower Egypt), Awlad Sheykh in Suhag governorate (Upper Egypt), and four villages in the oasis region rarely include these people. They are predominantly composed of couples and their children.

Household Cycle

One important reason for variations in household formation is the household cycle. Table 8 shows the household cycle in the 19 villages. Young couples start their married life in an independent household, or in their parent's household. However, it has become rare for the couples to live with their parents in the long term. In fact, household heads aged between 25 and 54 rarely live with their parents. However, when they become older, some couples move in with their married sons. Thus, the household composition shifts vertically with age.

The horizontal relationship between brothers follows this vertical parent–son relationship. As is shown in Table 8, the household heads tend to live with their brothers in the early stages of their marriage. This may happen when the young couple resides in their parents' household, which means that the cohabitation with a brother occurs as a by-product of the vertical son–parent relationship.

TABLE 8 *Types of household changes with age of the household heads (%)*

	15–24	25–34	35–44	45–54	55–64	65 & above	Total
Single person	9.9	1.3	0.7	1.5	5.3	16.7	4.2
Couple	19.8	12.5	2.2	2.1	7.4	25.8	8.6
Parents + unmarried children	27.0	65.7	78.0	72.7	59.3	30.2	63.6
Head + unmarried & divorced children	17.5	9.2	8.2	11.3	10.1	6.1	9.4
Couple + children + parent(s)	0.8	4.1	5.2	4.1	1.4	0.0	3.3
Other	25.0	7.2	5.7	8.3	16.5	21.1	11.0
Total	100.0	100.0	100.0	100.0	100.0	100.0	100.0

Source: 2005 Rural Household Survey data.

Marriages between Relatives

The most common pattern of marriage between relatives is marriage with a cousin on the father's side. In fact, the proportion of household heads married to their first cousin on their fathers' side (*awlad amm/amma*) is 37.3% in Beni Samrag in Minya governorate, 26.3% in Awlad Sheykh in Suhag governorate, and 24.0% in Sheykh Isa in Qena governorate. The total proportion of all marriages to relatives reaches 67.2% in Beni Samrag, 75.9% in Awlad Sheykh, and 60.6% in Sheykh Isa. Thus, the households in these villages prefer marriage to their relatives.

The majority of householders in Lower Egypt and the oasis region, on the other hand, marry villagers without kinship ties (Table 9). In the four villages in the oasis region, the proportion of householders married to villagers without a kinship tie is 35.8% in Munira, 47.2% in Bulaq in Kharga Oasis, 63.1% in Rashda, and 68.1% in Balat in Dakhla Oasis. Thus, people in the two villages in Dakhla Oasis tend to marry people without a kinship tie more than others in the Nile Valley or Kharga Oasis tend to do.

There are also household heads who marry people from outside the village. This is common in Abu Senita in Menufiya governorate (Lower Egypt), Kafr Shubrahur in Daqahliya governorate (Lower Egypt), and Kafr Beni Hilal and Sidi Oqba in Buheyra governorate (Lower Egypt). This phenomenon is related to the proximity to the large city and migration to it. In the case of Abu Senita, where 26.2% of household heads are married to someone from outside the village, there are many villagers who have migrated to Cairo. In the two villages in

TABLE 9 *Relationships of household heads to their spouses (percentage of total marriages) (2005)*

		Father's side first cousin	Mother's side first cousin	Father's side second cousin	Mother's side second cousin	Father's side relative	Mother's side relative	From same village	Same workplace	Other	Total	(Number)
Lower Egypt	Abu Senita	11.4	6.7	2.3	0.2	3.7	3.9	44.9	0.8	26.2	100	519
(Central)	Zahra	7.2	10.7	2.0	1.5	3.7	5.5	51.8	0.6	17.1	100	543
	Abu Tawala	7.2	7.8	1.2	2.2	5.2	7.2	52.7	0.2	16.5	100	503
Lower Egypt	Hamra	15.2	8.4	0.6	0.0	2.1	3.7	55.5	0.1	14.4	100	842
(Peripheral)	Kafr Shubrahur	7.3	8.1	0.0	0.8	2.4	2.4	6.5	0.8	71.8	100	124
	Kafr Beni Hilal	13.9	6.7	1.2	0.6	2.7	9.6	42.5	0.8	22.1	100	520
	Sidi Oqba	18.6	7.1	0.8	0.4	0.4	3.3	47.8	1.2	20.4	100	510
Middle Egypt	Homa	25.1	10.3	6.3	1.7	6.9	4.4	33.0	0.6	11.8	100	525
	Damashqin	10.2	8.2	2.9	0.2	5.1	4.7	54.3	1.4	13.1	100	512
	Beni Samrag	37.3	9.9	4.8	1.1	10.5	3.6	21.0	0.4	11.4	100	525
Upper Egypt	Kawamel Bahry	16.9	15.1	6.3	0.4	10.7	6.1	26.3	0.4	18.0	100	544
	Awlad Sheykh	26.3	18.1	9.8	3.5	14.9	3.3	17.1	0.0	7.0	100	543
	Sheykh Isa	24.0	10.8	5.6	1.1	8.7	10.4	22.7	0.4	16.2	100	462
Oasis	Munira	15.3	14.6	6.0	0.8	8.2	3.7	35.8	0.4	15.3	100	268
	Bulaq	10.4	10.8	2.4	1.6	6.8	3.6	47.2	2.0	15.2	100	250
	Rashda	6.6	5.4	1.5	0.2	8.1	6.2	63.1	0.4	8.5	100	518
	Balat	9.8	4.2	1.4	0.6	2.8	4.6	68.1	0.4	8.2	100	501
Total	Total	15.9	9.4	3.2	0.9	5.9	5.2	43.0	0.6	16.0	100	8,496

Source: 2005 Rural Household Survey data.

Buheyra governorate, where 22.1% and 20.4% of the household heads are married to someone from outside the village, there are many migrants to Alexandria. In the case of Kafr Shubrahur, this may be because this village was historically an *'izba.*[6]

Family Structures in Four Survey Villages

Our original surveys in 19 villages targeted households, which is the usual method of conducting a survey. These surveys provide rich information about social relationships within households—but not about families, defined as a social institution composed of households. Thus, we chose four villages from among the 19 surveyed villages to examine in detail the structures and socio-economic roles of families in village life.

Family (aila) of the households are classified based on the answers to the question asking *laqab al-'a'ila* (family name) of the household heads. To avoid the confusion of a family name with a household name (or, more precisely, the name of a household head), *laqab al-'a'ila* was used instead of *ism al-'a'ila. Ism al-'a'ila* also means family name, but it can be understood by respondents as asking their names. The collected *laqab al-'a'ila* were checked with the help of the informants in each village (such as CAPMAS staff or the umda) and revised when the response indicated a misunderstanding of the question.

The four villages are Abu Senita, Homa, Awlad Sheykh, and Rashda, located in Lower, Middle, and Upper Egypt, and the oasis region, respectively. These four villages are selected as typical of each region. We prepared a digital map and conducted interviews to collect information on how the villagers perceive family (aila).

Abu Senita is located in Bagur district (Markaz), Menufiya governorate, and it is a typical region with a peasant economy in Lower Egypt. It has a population of 4,468 (1996). Homa is in Wasta district, Beni Suef governorate, with a population of 6,148 (1996). It is located about 100 kilometres south of Cairo, near Wasta city. Homa and the governorate to which it belongs, Beni Suef, may be considered to be part of Middle Egypt, although it officially belongs to Upper Egypt. In terms of dialect and other cultural characteristics, its inhabitants are similar to the people of Upper Egypt, but they do not consider themselves to be Upper Egyptian (Ibrahim 1993).

---

6 *'Izba* historically signifies a small settlement of peasants who provided labour for farms in Lower Egypt. For the history of Kafr Shubrahur, see (Kato 1982).

Awlad Sheykh is located in the Dar al-Salam district, Suhag governorate, and it has a population of 5,242 (1996). The district to which the village belongs is known to have the highest rate of poverty in Egypt. The governorate is also known for its long history of migration to Cairo and Alexandria, as well as abroad, especially to the Gulf countries. Although far from the two internal cities (it is about 500 kilometres from Cairo), the governorate has historically been one of the largest sources of temporary workers, especially before World War II.

Rashda is located in Dakhla Oasis, in the New Valley governorate, and is studied in detail in the following chapters.

Size of Family

The typical size of a family, based on number of households, differs by village. In Abu Senita and Rashda, most families are composed of a small number of households (Table 10). In fact, 75.2% of the households in Abu Senita and 54.8% of those in Rashda belong to families composed of fewer than 30 households. On the other hand, Homa has larger families, many with 40 or more households. Of the households in Homa, 54.0% belong to families with 40 or more households.

Awlad Sheykh contrasts sharply with Abu Senita and Rashda. Of the households in Awlad Sheykh, 94.2% belong to families with more than 50 households. More than half the households belong to two families, each composed of more than 90 households: its largest family is that of the umda (village chief), with 232 households, and the second largest is the family of H, with 110 households. In contrast, the largest family in Abu Senita (the family of B) and Rashda (the family of AH), which are the families of the umda, have only 44 households (7.3% of 600 households) and 84 households (15.4% of 600 households), respectively.

Composition of Family

A difference in the composition of families is also found in the way that the villagers with whom we talked in Abu Senita and Awlad Sheykh described their kinship structure. In Awlad Sheykh, umda and other villagers called the largest unit of kinship aila *kabira* (large family) or *qabila* (tribe), and its subdivision *bait*. In Abu Senita, on the other hand, the villagers unanimously explained that there was no subdivision within aila and that an aila is normally composed of three generations; that is, an individual's household, and his brother's, father's, and grandfather's households.

Households whose aila are unknown, or who are originally from outside the village, are rare in Awlad Sheykh and Rashda. In contrast, Abu Senita and

TABLE 10 *Size of the aila according to the number of households (number of the aila and households)*

Number of households per aila	Abu Senita		Homa		Awlad Sheykh		Rashda	
	Aila	House-hold	Aila	House-hold	Aila	House-hold	Aila	House-hold
1/4	42	82	13	28	1	2	16	30
5/9	11	84	4	29	1	7	7	50
10/19	8	112	7	71	0	0	5	97
20/29	7	174	3	78	1	26	3	71
30/39	1	38	1	32	0	0	0	0
40/49	2	88	2	83	0	0	1	42
50-	0	0	4	241	5	564	1	84
From outside	6	7	14	19	0	0	39	172
Unknown	14	16	18	19	1	1	0	0
Total	91	601	66	600	9	600	72	546

Notes: 1. Households are classified by aila according to the response to the question item: "What is your family name (*laqab al- 'a'ila*)?"; 2. "From outside" refers to the names of aila that were originally from outside the village. "Unknown" refers to the names of aila that were originally from the village but we were unable to determine to which aila they belonged.
Source: 2005 Rural Household survey data.

Homa have many villagers whose aila is unknown. Of 91 aila, 13% are unknown in Abu Senita.

Aila that is perceived by the villagers as originally from outside the village have a significant presence in Rashda. Thirty-nine of 72 aila are originally from outside the village; 31.5% of the households belong to these aila.

Residential Pattern of Family

To understand the importance of the family as a social tie among households, it is necessary to understand the residential pattern of the household members of the same family, because social life is held in that space.

In the four villages, households of the same aila seem to live side by side (Map 5.1, 5.2, 5.3, 5.4[7]). This should be a natural phenomenon, because sons

7 The household survey covers 550 households. Most of the old families live in the oldest area to the south-west, which was not included in the household survey. The fieldwork was conducted in 2005.

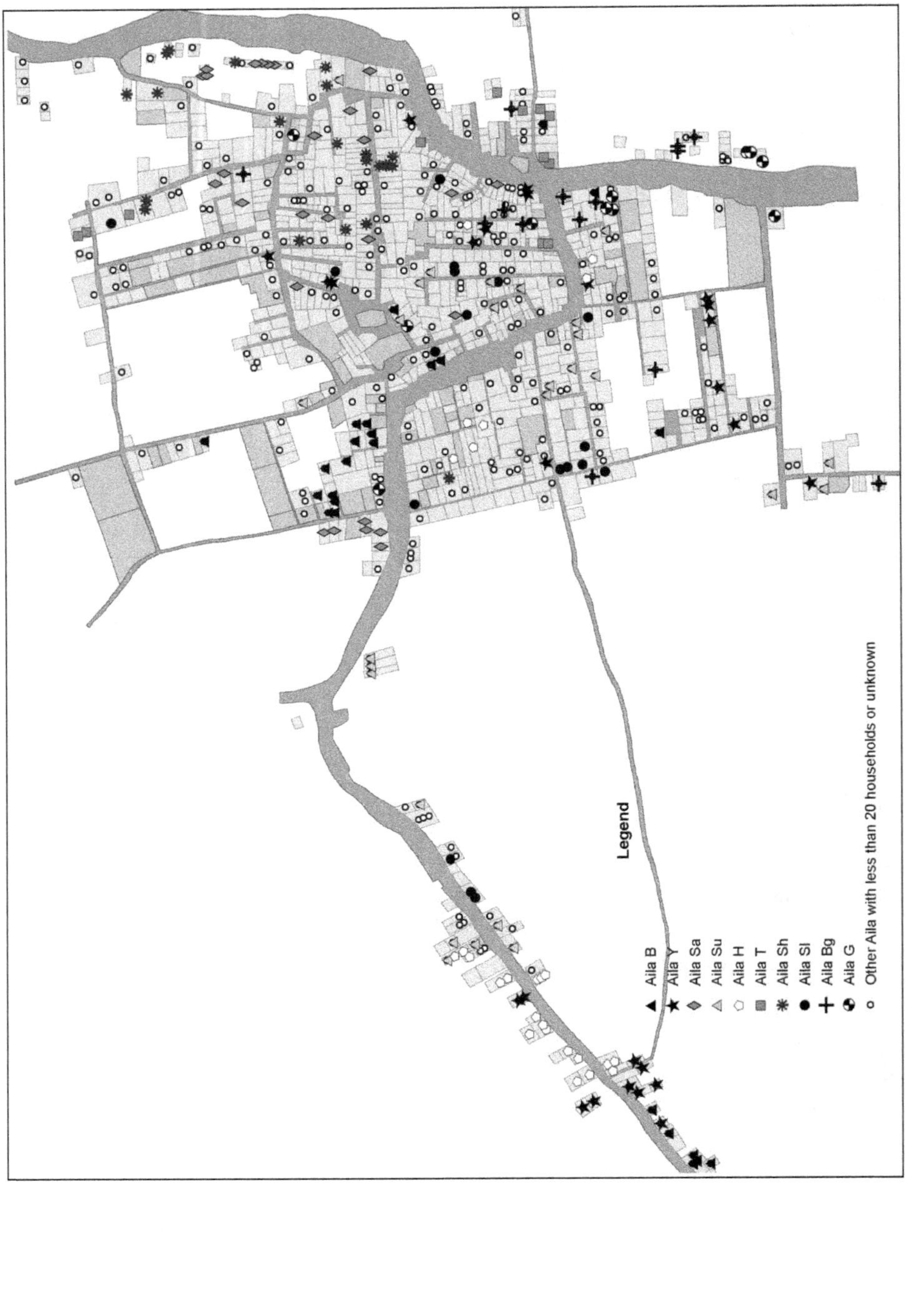
Legend
Aila B
Aila Y
Aila Sa
Aila Su
Aila H
Aila T
Aila Sh
Aila Sl
Aila Bg
Aila G
Other Aila with less than 20 households or unknown

MAP 5.2 *Aila of household heads (Homa). Source:* 2005 Rural Household Survey data.

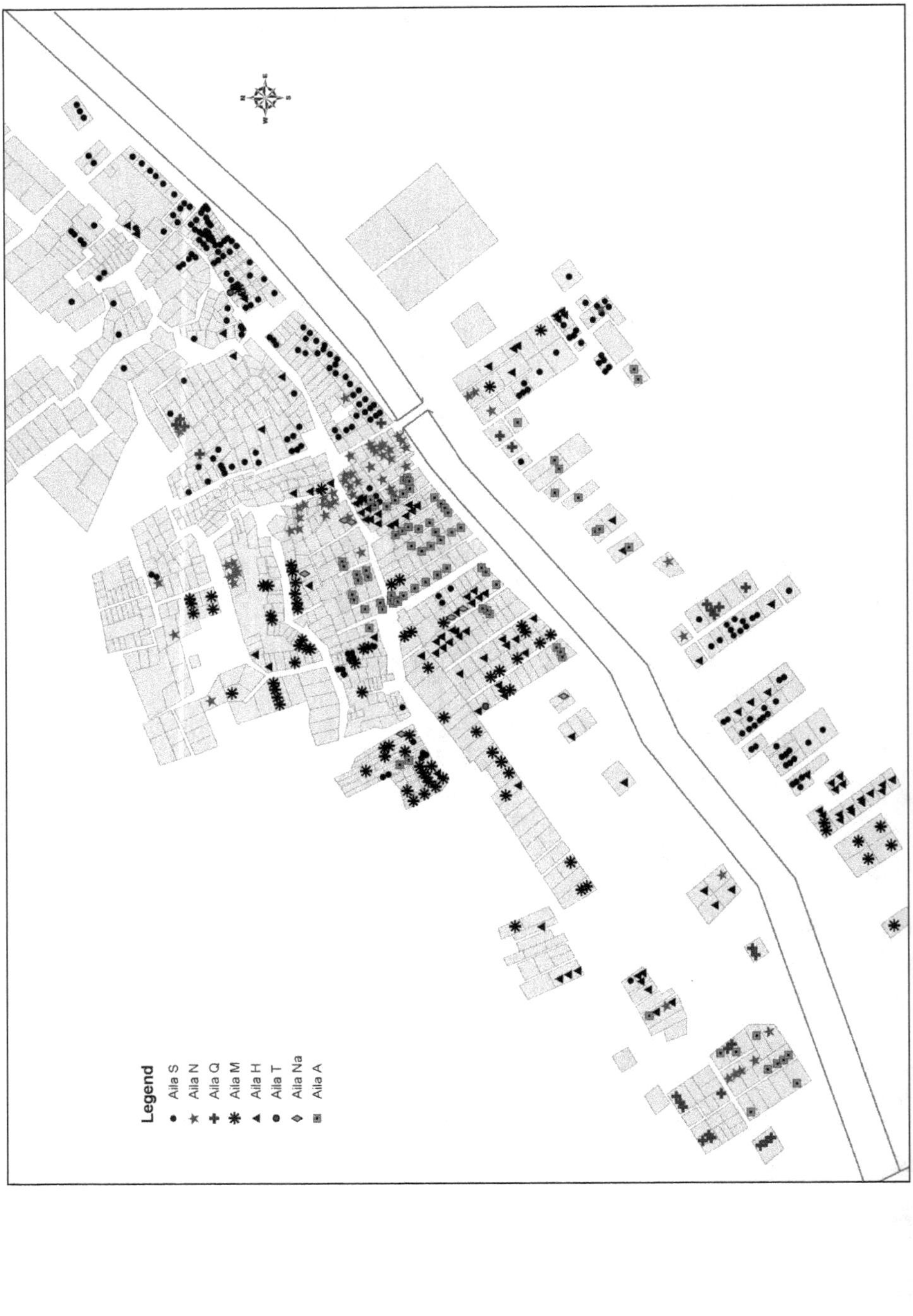
Legend
Aila S
Aila N
Aila Q
Aila M
Aila H
Aila T
Aila Na
Aila A

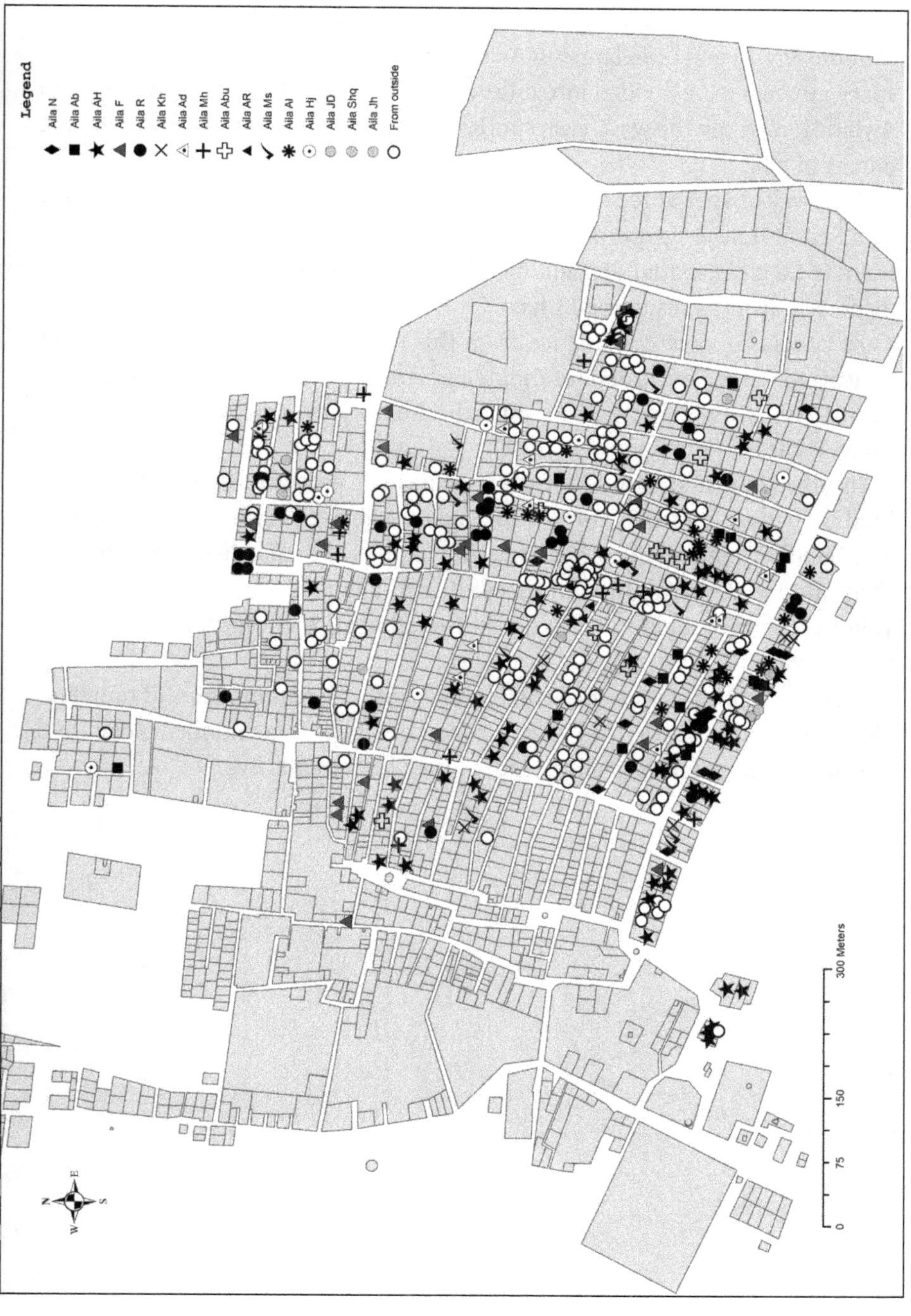

MAP 5.4 *Aila of household heads* (*Rashda*). *Source*: 2005 Rural Household Survey data.

may build their houses on inherited land. What seems surprising is the difference in the visibility of the spatial division by aila.

As the number of households belonging to the same aila is larger in Homa (Middle Egypt) and even larger in Awlad Sheykh (Upper Egypt), the residential space appears to be divided into different aila quarters. Indeed, the villagers in Awlad Sheykh call the west, east, south, and central areas of their village by the names of aila.

This way of spatial recognition is almost completely absent in present-day Abu Senita (Lower Egypt) and Rashda (the oasis region). In these two regions, there is no clear spatial division by aila. All the aila are small in size and are dispersed over the residential space.

In conclusion, we can deduce from this examination an important point relating to the structure of households and families in Egypt, which is that family structures differ by region, while the household structure is almost the same, regardless of region.

In Abu Senita and Rashda, an aila appears to be perceived as an intra-household tie between a small number of households. In the two villages of Upper and Middle Egypt, on the other hand, the frequency of marriage between relatives and clear spatial divisions according to aila lead us to believe that the villagers perceive aila as beyond the relationships of individuals around households.

It seems that this difference corresponds with the relationships of individuals around households, and it is a difference in the economic, social, and political circumstances in which individuals/households live.

CHAPTER 4

Social Economy

In Chapter 3, we confirmed that regional differences existed in terms of family structure, but they were not observed in household structure. Moreover, we suggested how such differences may be identified. The reason for the differences is that the family has different roles from the household in both social and economic life. This means that regional differences in family structure reflect social and economic circumstances. This chapter attempts to explain this result at the village level, using the agglomerated statistical data produced by the Egyptian government and the micro data and information on 19 villages mentioned above collected through our surveys, and to locate the villages of Dakhla Oasis within the Egyptian social economy.

Social and Economic Structures in Egypt

The three indicators of regional differences in the social economy in Egypt are income level, educational level, and employment structure. The indicators used here are limited to these three, based on the assumption that these are the most basic and important elements of the Egyptian society and economy.

Income level differs remarkably between Greater Cairo and other regions. Greater Cairo has the highest levels. Upper Egypt has the lowest income level in both urban and rural areas. Income levels in the Frontier Governorates to which the governorate of the New Valley belongs are higher than those in other urban governorates. Rural areas of the Frontier Governorates also have higher income levels than those of Lower and Upper Egypt.

Educational level follows the same patterns as income level. Almost all research points to the importance of education for social and economic development. Educational level is clearly one of the key factors in Egyptian society, and this has a clearly regional dimension.

Greater Cairo has the highest proportion of university graduates, followed by other urban governorates, then urban Lower and Upper Egypt. Urban areas of the Frontier Governorates also have high percentages of university graduates. Rural areas, by contrast, have higher rates of illiteracy than urban areas, with the Frontier Governorates and Lower Egypt having lower rates of illiteracy than Upper Egypt.

 | DOI 10.1163/9789004317390_006

As to the governorate of the New Valley, Kharga and Dakhla Oases have remarkably high levels of education, compared with those in the Nile Valley. The illiteracy rates in Kharga, Dakhla, and Farafra Oases are 7.6%, 11.9%, and 5.4%, respectively, compared with the national average of 26.6% in 2006.

Urban areas have much higher percentages of people in paid employment in the government sector. There are some differences within urban areas. In terms of industry, Greater Cairo and other urban governorates have more manufacturing, whereas Lower and Upper Egypt, and especially the Frontier Governorates, have higher percentages of people employed in public administration/defence and education, which are government-related activities.

In rural areas, agricultural employment predominates, with some regional differences: Upper Egypt has the highest, and the Frontier governorates have the lowest percentage of employment in agriculture. Non-farm employment is mostly provided by the government sector.

However, the data on these indicators suggest that the spatial variation is not that simple. Mapping the data at the smallest administrative unit (towns/villages) indicates considerable variations in income and educational level within each region of Lower and Upper Egypt, the Frontier governorates (Map 6).

Regional Categorization of Social and Economic Structures

Cluster Analysis

After the overview of the regional differences in social economic structures, a regional categorization was conducted based on the two kinds of agglomerated data at the smallest administrative unit (town/village) from a population census in 1996 and the Household Income & Expenditure Survey in 1999–2000. The regional diversity of Egyptian society is clarified using a cluster analysis of income, employment, and educational level.

Cluster analysis is a statistical method for grouping objects into categories or clusters according to the similarity of their attributes. It should be noted that its results neither indicate any causality nor identify which of the attributes has the greatest affect. Cluster analysis is used despite these shortcomings because it is consistent with the aim of this study, which is to create a regional socio-economic categorization without predicting any outcome, so as to avoid a stereotypical regional division. The advantage of cluster analysis is that objects are classified without a priori information from outside sources. It is an exploratory method, unlike many other statistical methods. It classifies objects without determining the number and definition of groups or classification

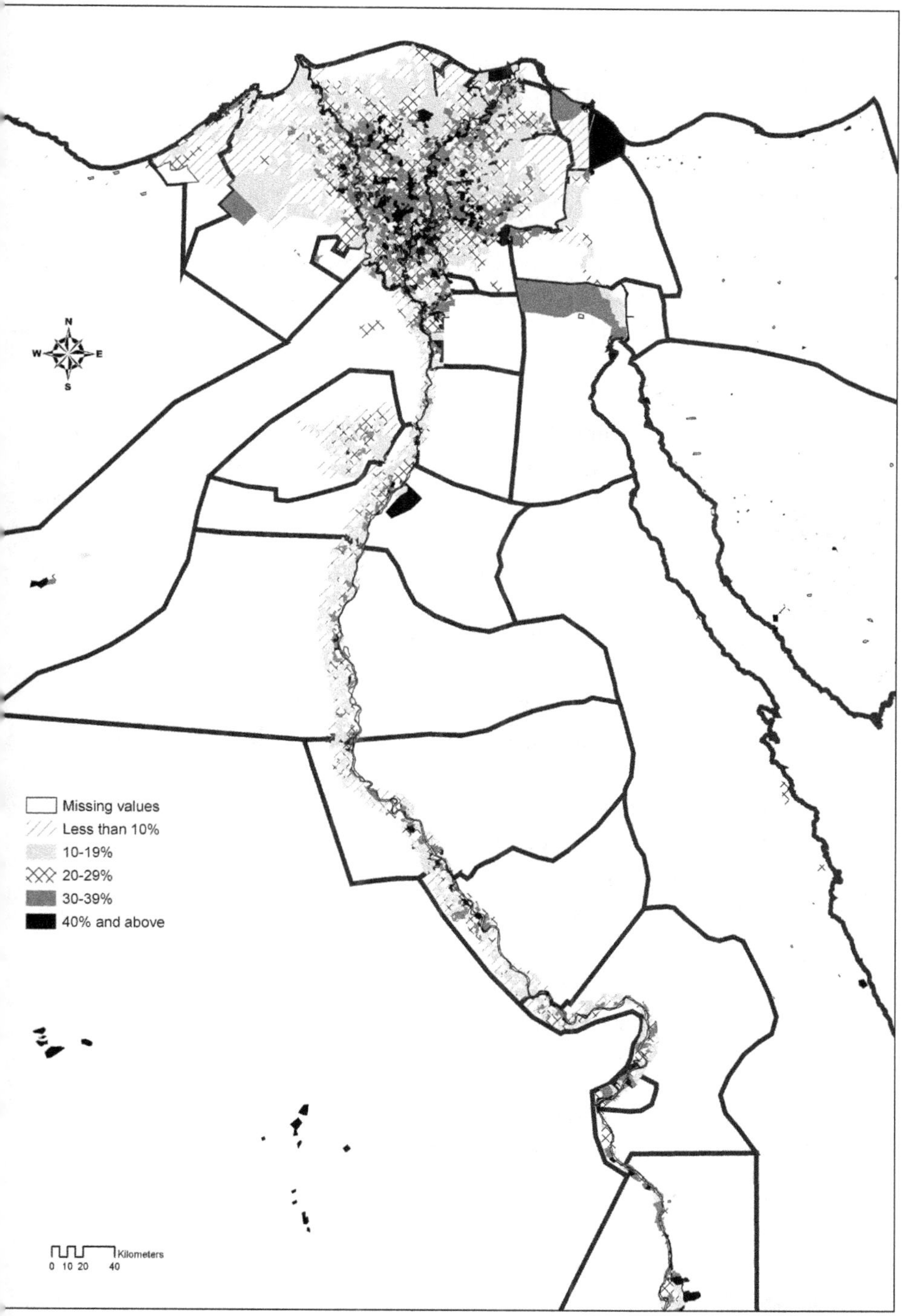

MAP 6 *Percentage of government employees (Percentage among people aged 15 years & older) (unit: town/village) (1996). Source*: Kato and Iwasaki 2008: 13.

criteria, using the mathematical method of calculating the similarity or dissimilarity among groups.

The analytical procedure was as follows. First, factor analysis was conducted to summarize the information contained in the original set of variables.[1] After the factor analysis, cluster analysis was performed to classify 5,345 units (town/village) covering all of Egypt using factor scores computed from the factor loadings for each town/village.[2]

A total of 41 variables were used for the analysis.[3] All variables are related to the indicators of income, educational level, and employment that are available from the two survey datasets.

Regional Categorization

Cluster analysis was conducted using the scores of the 10 factors obtained from the factor analysis. It produced seven clusters characterized by similar features of income, employment, and educational level (Table 11). A detailed description of these seven clusters can be found below and can be further understood by examining Map 7, which shows the geographical distribution of these clusters.

The findings can be summarized as follows.

1. Cairo is the primary city in terms of the urban functions of administration/commerce/industry, and in terms of size, as shown in the spatial distribution of Cluster 1, "predominance of industrial workers", and Cluster 6, "mixture of government employment, commercial and industrial activities". Moreover, as shown in the spatial distribution of Cluster 7, "high income level and dominated by service industry", the urban rich class is exclusively found in Cairo. Thus, Cairo can be identified as the pre-eminent centre for administration and economy.

---

1 Factor analysis is a method of reducing large numbers of variables into fewer unobserved variables called factors. The three indicators used in the analysis are expected to correlate highly with each other and may not converge into adequate cluster numbers. Therefore, factor analysis (principal component method) was undertaken to summarize the information contained in the original set of variables. Then the extracted factors are rotated using the varimax method. This method has the advantage of yielding the solutions that are easiest to interpret. Each factor is interpreted according to its loading value.

2 Our decision on the number of factors to retain was based on the eigenvalue, where the amount of variation explained by each factor must be greater than 1. Therefore, the first 10 factors with eigenvalues greater than 1 were retained. These 10 factors explain 74% of the total variance of the original variables.

3 For details, see Kato and Iwasaki 2008: 11.

TABLE 11 *Regional categorization according to the cluster analysis*

Cluster	Geographical location	Characteristics
1	Suburbs of large cities (Cairo, Alexandria, Port Said) Suburbs of Mahalla Kubra city "Provincial cities in the central and southern parts of Lower Egypt (governorates of Menufiya, Sharqiya, Gharbiya, Qalyubiya)" Villages in the southern parts of Upper Egypt	Predominance of industrial workers
2	Villages in the suburbs of provincial cities, especially in the central and southern parts of Lower Egypt (Daqahliya, Menufiya, Sharqiya, Qalyubiya) Villages in the southern parts of Upper Egypt, and oases in the Frontier governorates	Markedly low ranked government employment
3	Villages in the northern parts of Lower Egypt (Daqahliya, Buheyra, Sharqiya) "Villages in Fayum governorate and other southern parts of Upper Egypt (Asyut, Suhag governorates)"	Predominance of agricultural self-employment
4	"Villages in the southern parts of Upper Egypt (Minya governorate and northern parts of Asyut governorate)"	Predominance of waged agricultural labour and low income
5	"Villages in the northern parts of Lower Egypt (governorates of Buheyra, Kafr Sheykh, Daqahliya)" Villages in Minya governorate	Predominance of large-scale agricultural enterprises
6	Large cities, and most of the provincial cities	Mixture of government employment, commercial and industrial activities
7	Some districts (*qism*) in Cairo	High incomes and dominated by service industries

Note: For details, see Kato and Iwasaki 2008: 14.
Source: CAPMAS, 1996 Population Census dataset, Household Income & Expenditure Survey 1999/2000 dataset.

2. Most of the provincial cities belong to Cluster 6, "mixture of government employment, commercial and industrial activities", except Mahalla Kubra and Damietta city. The main feature of this cluster is the predominance of government sector employment. This fact led us to suppose that while the private-led economic development since the

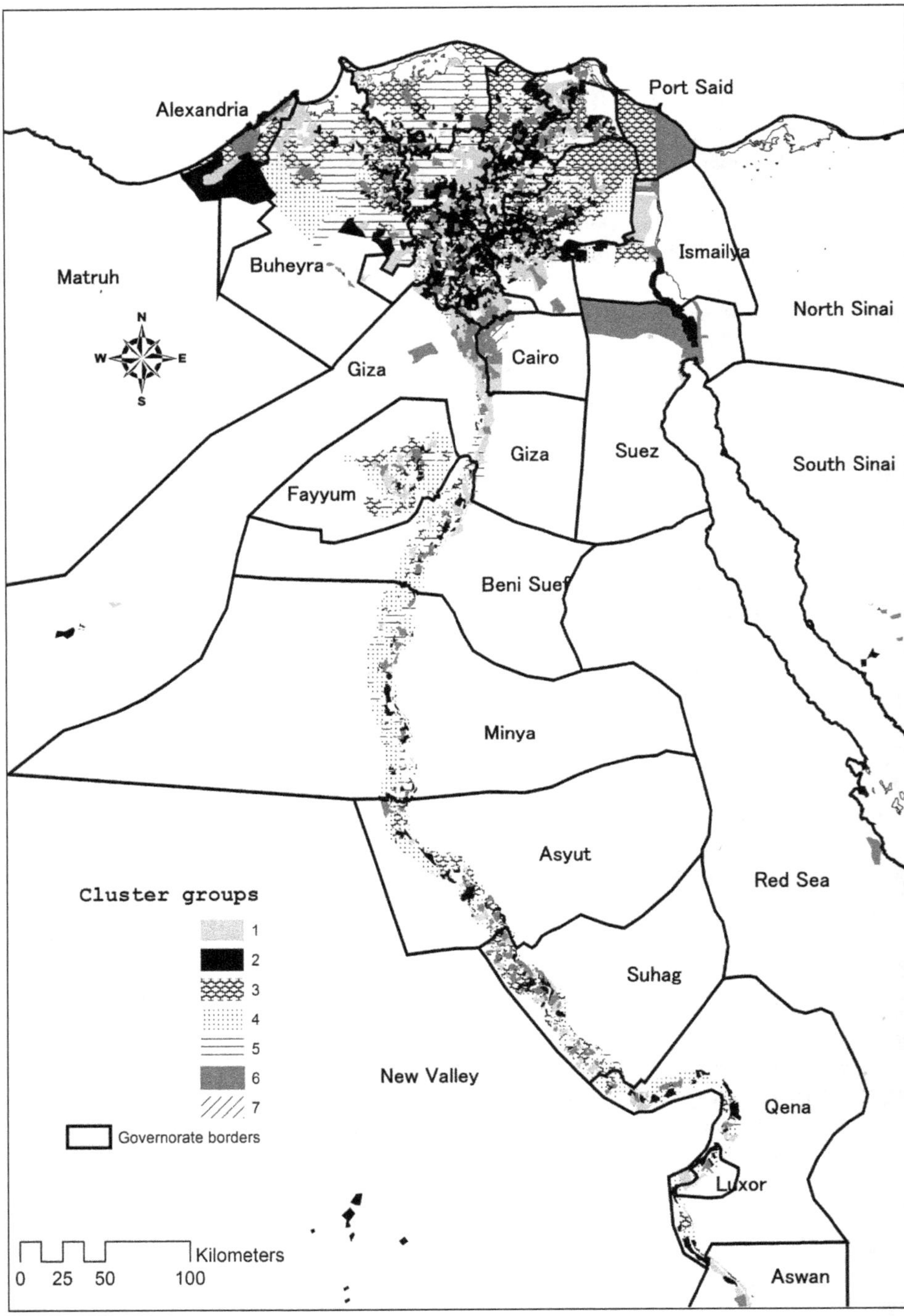

MAP 7 Result of cluster analysis (unit: town/village) (1996, 1999/2000). *Source*: Kato and Iwasaki 2008: 15.

beginning of Open-door policy (*Infitah*) took place in Cairo, provincial cities still depend upon the government-led economy formed in the period of the planned economy in the 1960s.

3. Unlike provincial cities, which seem to be homogenous regardless of their geographical location, rural areas vary widely and qualitatively across the commonly used division of Lower and Upper Egypt. These regions are as follows.
   1. Central and southern Lower Egypt and the Frontier governorates, which belong mainly to Cluster 2, "markedly low rank government employment", are characterized by the relatively high educational level and predominance of government sector employment.
   2. Southern Upper Egypt from Suhag to Aswan governorates mainly belongs to Cluster 1, "predominance of industrial workers". The residents there seem to depend upon non-agricultural employment opportunities in local factories and, through migration, on the distant large cities.
   3. Kafr Sheykh and Buheyra governorates belong to Cluster 5, "predominance of large-scale agricultural enterprises".
   4. A majority of villages in southern Upper Egypt, in places such as Minya governorate, belong to Cluster 4, "predominance of waged agricultural labour and low income". In these areas, poverty seems to be associated with the agricultural sector.

Description of 19 Villages Using Basic Indicators

In this section, we examine the characteristic of 19 villages in greater detail according to the three basic indicators: education, income, and employment.

Education

Illiteracy rates, which average 35% in these 19 villages, rise to more than 50% in the villages of Middle and Upper Egypt except for Sheykh Isa in Qena governorate. On the other hand, the oasis villages, which are the most remote of the 19 villages, have less than 20% illiteracy among those aged 10 years and above.

Income

However, low educational level does not necessarily mean that the living standard of the village is low or poor. Let us consider household income. Household income is defined as the sum of income from 1. farming (including) self-employment, 2. wage employment, 3. non-farm self-employment, 4. real estate,

5. financial assets, and 6. transfers earned during the 12 months prior to this fieldwork (2003–2004 for six villages and 2005 for 13 villages).[4]

At the time of the household survey in 2005, the average per capita annual household income of all the 19 villages was 2,336 Egyptian Pounds. The villages with the lowest household income were the three villages of Middle Egypt followed by Abu Senita in central Lower Egypt, and the two villages of peripheral Lower Egypt, Hamra and Borr Bahry. The villages with the highest household income, on the other hand, are Awlad Sheykh and Sheykh Isa villages in Southern Upper Egypt, followed by Sidi Oqba in peripheral Lower Egypt, and Munira and Rashda villages in the oasis region. Thus, focusing on the village level gives a different picture than the widely known fact that rural Upper Egypt is the poorest region.

The primary income source is non-farm wage employment for all the villages. It provides around 40% of household income. The proportion of farm income is only 30% to 40%, except for Liwa Subih and Hamra villages.

Transfer income is an important source of income and provides more than 10% of total household income in every village. The households in Awlad Sheykh and Sheykh Isa villages, notably, receive a remarkably high proportion of transfers. This is because Awlad Sheykh has many migrants to Abu Dhabi in the UAE, and Sheykh Isa has many migrants to Harghada city on the Red Sea

4 Total household income is defined as the sum of farm income, wage income, non-agricultural self-employment income, and other sources of income that households received during the 12 months prior to the survey. The definition of household income in our survey follows that of CAPMAS. According to CAPMAS, household income comprises the following six components. (A) Farm income is defined as gross agricultural revenue minus material costs. Gross agricultural revenue is the gross value of cash derived from agricultural activities including cultivation, livestock husbandry, forestry, and fishery. It includes the value of the agricultural products consumed by the household. (B) Non-agricultural wage income is defined as the sum of wages, bonuses, and allowances received from employers such as governmental institutions and public and private enterprises. In this paper, wage income is classified into wage income received from non-agricultural activities and that from agricultural activities. (C) Non-agricultural self-employment income is the total value of net cash income (sales minus material costs) from non-agricultural self-employment activities in commerce, manufacturing, construction, transportation, services, etc. (D) Income from real estate includes the rental income from land, apartments or buildings, and other real estate. It also includes imputed rent (the estimated value of the property in cases where a household rents out its house). (E) Financial income is the income derived from bonds, deposit interest from banks, post offices, "investment certificates", etc. (F) Other income is mainly transfer income comprising retirement pensions, social security, social welfare payments to poor families, and private transfers from within Egypt or from abroad (gifts from relatives and money sent by migrants). In most cases, other income is from retirement pensions.

TABLE 12 Household income distribution and income sources in 19 villages (2005)

		Overall household income (LE)		Income sources (%)								Gini coefficient
		(per capita)	(per household)	Farm	Farm wage	Non-farm wage	Non-farm self-employment	Real estates	Finance	Transfer	Total	(per capita)
Central Lower Egypt	Abu Senita	1,829	8,801	33.9	6.6	29.9	6.8	15.3	0.2	7.3	100.0	0.311
	Kafr Shubrahur	2,551	9,880	21.1	1.6	40.1	12.3	8.4	0.9	15.6	100.0	0.299
Peripheral Lower Egypt	Zahra	2,410	8,773	42.3	2.9	27.7	3.1	13.6	0.1	10.3	100.0	0.291
	Abu Tawala	2,331	7,790	28.1	2.0	31.8	4.2	17.1	0.9	15.9	100.0	0.283
	Kafr Beni Hilal	2,627	9,599	21.7	3.7	38.6	5.7	12.5	0.9	16.9	100.0	0.349
	Sidi Oqba	2,853	11,635	46.3	13.3	13.9	5.2	11.2	0.0	10.1	100.0	0.305
	Hamra	2,140	10,507	72.6	0.0	10.6	6.3	6.7	0.1	3.7	100.0	0.243
	Borr Bahry	1,903	11,091	18.9	1.4	43.9	5.5	13.0	0.3	17.1	100.0	0.285
Middle Egypt	Homa	1,475	6,744	42.5	6.2	25.8	4.8	11.3	0.7	8.8	100.0	0.312
	Damashqin	1,789	8,619	29.8	7.3	32.9	4.0	13.4	0.2	12.4	100.0	0.292
	Beni Samrag	1,847	9,870	36.8	12.0	18.6	5.4	8.3	0.6	18.3	100.0	0.355
Southern Upper Egypt	Kawamel Bahry	2,143	9,099	45.7	9.3	19.0	8.2	9.1	0.1	8.6	100.0	0.369
	Awlad Sheykh	3,091	10,684	27.5	10.1	12.7	4.7	7.0	0.2	37.9	100.0	0.485
	Sheykh Isa	2,985	14,757	18.7	1.8	41.3	4.0	10.3	0.3	23.7	100.0	0.412
Oasis	Munira	2,701	10,813	18.8	1.3	57.3	2.2	2.7	0.1	17.5	100.0	0.333
	Bulaq	2,422	9,726	16.5	3.8	52.1	5.4	2.1	0.3	19.9	100.0	0.306
	Rashda	2,707	10,520	29.7	1.8	48.4	4.4	5.9	0.0	9.8	100.0	0.308
	Balat	2,655	10,446	32.4	1.0	43.9	2.5	6.3	0.1	13.8	100.0	0.293
	Liwa Subih	2,433	11,381	77.4	8.3	11.3	0.0	1.4	0.0	1.5	100.0	0.313
Total		2,336	9,935	32.7	5.4	31.4	5.0	10.5	0.3	14.6	100.0	0.349

Source: 2005 Rural Household Survey data.

coast. Transfer income also provides more than 15% of income in Abu Tawala and Kafr Beni Hilal in peripheral Lower Egypt.

In the two villages of Awlad Sheykh and Sheykh Isa, the income disparity among the villagers as expressed in the Gini coefficient is remarkably high. This is because of transfer income, which provides a large income to the households. Beside the villages with migration, the income disparity inside a village is small. The village with the lowest income disparity is Hamra in peripheral Lower Egypt, followed by Abu Tawala in central Lower Egypt and Borr Bahry in peripheral Lower Egypt. Other villages have similar levels of income disparity between villagers.

Employment

Non-farm activities are one source of diversity. This is because each village is linked to different labour markets, and its access is determined not only by human capital (education) but also by the networks that the villagers have constructed between their village and the workplace.

In today's rural Egypt, non-farm employment is the major source of employment. If households are classified by the number of days worked in farming and non-farm sectors, practically all the households are engaged in non-farm activities. Households in the villages in the oasis region are especially highly dependent on non-farm activities for employment.

The main source of non-farm employment is the government sector. In every village, being a public servant in the local unit or public institutions and being a schoolteacher are popular jobs. On average, half of the non-farm employment is provided by the government sector. However, its importance differs by village. It is very important in villages in central Lower Egypt, such as Kafr Shubrahur, Abu Tawala, Abu Senita, and Kafr Beni Hilal. In oasis villages, the proportion of people employed by the government is surprisingly high: 80% in Bulaq, Rashda, and Balat villages. Thus, the economic life in the oasis region is highly dependent on the government, although it is far from Cairo.

In contrast, Hamra, Homa, Beni Samrag, and Kawamel Bahary villages have high proportions of non-farm private sector wage workers. Non-farm workers in Hamra village are workers in commerce and transportation in nearby towns and Cairo. Non-farm workers in Beni Samrag are construction workers in Cairo and nearby towns. Those in Kawamel Bahary are merchants in the wholesale vegetable market in Cairo.

Personalities of the 19 Villages

The regional categories resulting from the cluster analysis in Table 4–1 are generally applicable to all 19 villages. However, there are a number of exceptions, because each village has its own circumstances and endowments determined by space and time. The socio-economic life of a village is not a result of simple causality of its endowments. It is an interaction between the endowments—natural and human capital, and the external conditions or connectivity—that form the characteristics peculiar to each village.

We call the characteristics peculiar to the village the "personality" of the village. For the formation of the personality of a village, its history is very important because the natural, economic, social, and political circumstances in which individuals/households live differ between and within the regions, and change over time. The personality of each village is formed in history. Table 13 shows the personalities of 19 villages according to basic indicators, although these personalities are no longer clearly observable in today's Egypt.

The personalities of the 19 villages may be summarized in the following four points, in terms of the above discussion. First, Borr Bahry in Kafr Sheykh governorate and Liwa Subih in the Nile Valley governorate are exceptions. In fact, Borr Bahry is a fishermen's village, and the latter village was newly created in 1985. Second, the villages in the oasis region are classified as one group. Except for Liwa Subih, four villages are characterized by government sector employment and high educational levels. Munira is a village founded by new settlements in the 1960s. However, it has similar characteristics to Rashda, Bulaq, and Balat villages. This is because it is part of the government's New Valley project.

Third, the four villages in Upper Egypt, including those of Middle Egypt and Southern Upper Egypt, have similar characteristics in terms of their high dependence on agriculture and migration, and their low educational level, although the period of village formation and the influence of Arab culture are different. Damashqin and Sheykh Isa are exceptions for reasons probably related to their relations with the outside. Damashqin is located in Fayum basin far from the Nile River but has close connections with Cairo because of the development of a road. Sheykh Isa is near Qena, capital of Qena governorate and has many migrants to Harghada on the coast of the Red Sea.

Fourth, the villages in Lower Egypt have different characteristics from those in Upper Egypt, except for Borr Bahry, a fishermen's village. At the same time, there are differences in villages between the central and peripheral areas. The four villages of Abu Senita, Kafr Shubrahur, Kafr Beni Hilal, and Abu Tawala are characterized by government employment, whereas the three villages of Hamra, Sidi Oqba, and Zahra are highly dependent upon agriculture. The reason for this difference is in the development pattern. The villages in central Lower Egypt developed based on labour-intensive agriculture and have

TABLE 13 *Personality of 19 villages*

Region	Village	Governorate	Population (2006)	Period of village formation	Village type	Land reform	Presence of umda	Presence of police station	Presence of 'arab	Location
Central Lower Egypt	Abu Senita	Menufiya	4,468	Old	qarya	×	○	×	×	6km from Bagur
	Kafr Shubrahur	Daqahliya	515	New	izba	○	○	×	×	8km from Sinbilawin
Peripheral Lower Egypt	Zahra	Sharqiya	5,082	New	qarya & izba	○	○?	×	×	15km from Zagazig & Bilbeis
	Abu Tawala	Sharqiya	5,458	Old	qarya	×	○	×	×	8km from Minya al-Qamh
	Kafr Beni Hilal	Buheyra	5,933	New	qarya & izba	○	○	×	○	5km from Damanhur
	Sidi Oqba	Buheyra	20,429	Old	qarya & izba	○	○	○(1975)	×	9km from Mahmudiya
	Hamra	Kafr Sheykh	9,252	Old	qarya & izba	○	○	○(1968)	×	7km from Kafr Sheykh
	Borr Bahry	Kafr Sheykh	8,214	New (1998)	qarya	×	×	○(2002)	×	10km from Biltim
Middle Egypt	Homa	Beni Suef	6,148	Old	qarya & izba	○	×	○(1952)	×	4km from Wasta
	Damashqin	Fayyum	6,501	Old	qarya	○?	○	×	×	14km from Fayyum

Southern Upper Egypt	Beni Samrag	Minya	9,810	Old	qarya & izba	○?	○	×	○	20km from Minya
	Kawamel Bahry	Suhag	4,461	New	qarya	○?	○	×	○	12km from Suhag
	Awlad Sheykh	Suhag	5,242	New	qarya	×	○	×	×	20km from Dar al-Salam
	Sheykh Isa	Qena	5,651	New	qarya	○?	○	×	○	8km from Qena
Oasis	Bulaq	New Valley	1,825	Old	qarya	×	○	×	×	25km from Kharga
	Munira	New Valley	1,915	New (1966)	qarya	×	×	○(2003)	×	22km from Kharga
	Balat	New Valley	4,273	Old	qarya	×	○	×	×	27km from Mut
	Rashda	New Valley	4,364	New	qarya	×	○	×	×	10km from Mut
	Liwa Subih	New Valley	2,710	New (1985)	qarya	×	×	×	×	

Notes: The basic indicators in the table are selected for the following reasons. 1. "Period of village formation" and "Village type" indicate whether the village is an old and traditional one that existed before the 19th century, or a new and modern one formed in the 19th century. The terms of old and new are used in *al-Qamus al-jughrafi* (Ramzi 1968). The old and traditional village and the new and modern village are called *qarya* and *'izba* in Arabic respectively; 2. "Land reform" indicates whether land reform was executed in the village in the 1960s, which indicates the influence of large landholdings before the land reform in the village; 3. "Presence of the umda and a police station" indicate the centralization of village politics. In today's rural Egypt, the traditional power and authority of the umda have rapidly decreased; some villages lack the post of the umda and have a police station; 4. "Presence of 'arab'" is the indicator to measure the influence of Arab culture in the village; 5. "Location" is the indicator to show the distance of the village from the local city that is the administrative centre in the region.

Source: Kato and Iwasaki 2011:148.

recently shifted towards non-agricultural sector employment for income generation, along with land fragmentation. The villages in peripheral Lower Egypt, on the other hand, were created and developed in modern times, by capital-intensive agriculture. A typical case is Sidi Oqba, located in an area reclaimed in the early 20th century, and composed of 57 small settlements (*ʿizba*).

Oasis Villages in Egyptian Village Map

We conclude the discussion in Part 1 by pointing out the characteristics of oasis villages, comparing them with other villages in the Nile Valley.

First, as mentioned above, we must confirm that the villages in the oasis region can be classified as a group. This is probably because the oasis villages share the characteristics discussed in Section two of Chapter 1, "Islands in the sand". It is a social space with its own cultural traditions, surrounded by desert. In contrast to the villages of the Nile Valley, life there depends on water. In oasis society, the land has no value if there is no water. However, as for islands in the sea, the oasis villages cannot survive without trade with the outer world. In fact, the inhabitants of the oasis come from outside and have diverse origins. The traditional oasis villages have functioned as a nexus of a trade network in the desert. They are a "fort" in the desert, surrounded by date palms and walls. In the pre-modern era, the desert was a dangerous space where the Bedouins thrived. Therefore, cultivation rarely extended outside the wall.

However, the circumstances have largely changed. The oasis region has been incorporated into the territory of a modern state and into the administration system as it formed. The Bedouins were obliged to settle in the villages by the settlement policy. As a result, the desert became a safe place. At the same time, the populations of villages grew by natural and social increase, which necessitated the expansion of land outside the wall to meet the food requirements of the increasing population. Thus, the land expanded outside the wall, and the landscape of villages changed.

The national desert development project added to this. In the past, the inhabitants obtained water from springs or shallow wells. However, the increased population and the national development project raised the demand for water. Thus, deep wells had to be drilled.

Drilling a deep well requires capital and technology so that the inhabitants must depend upon the government and enterprises. The result is the paradoxical situation in which the oasis villages that are remote from the centre and have little connection with it share the characteristics of government sector employment and high educational levels with the villages of central Lower Egypt, which have developed as agricultural areas for a long time. With different initial conditions and itinerary, they reached the same result.

PART 2

Rashda in History

CHAPTER 5

Formation of Rashda

Rashda is now the second biggest village by population after Mut, the capital of the Dakhla Oasis District. However, Rashda in not an old village, but a new one, formed by the migrants from outside since the beginning of the 19th century. The rapid increase of population suggests the relative abundance of water in Rashda. This chapter aims to follow the history of Rashda from the start of migration to the Rashda area at the end of the 18th century until the formation of the Rashda settlement, and to reconstruct "Old Rashda", based on local documents and the results of the field surveys. In the 19th century, Rashda belonged administratively to Qalamun Village, the administrative centre of the region at that time.

Rashda at the End of the 18th Century

The name of Rashda village comes from the Roman spring called Rashda, which was located at the northern tip of the sand-hill on the western side of present-day Rashda. It has now dried up.

Although its name can be traced back to ancient times, and it is now the second most populous village in Dakhla, the present Rashda is a "new" village. It was formed in the modern age, after the 19th century. A series of local documents owned by local notables make it possible to trace the formation of the village of Rashda.[1]

We begin the discussion with a document dated at the end of the 18th century. This is a document dated 12 Rajab 1206 (1792) concerning the debt of a farmer. The document states as follows.

> The undersigned, two honourable gentlemen of full age and capacity, Mansur Farraj Sayyidin and Miftah Awd Abd al-Bari, originating from the people of the village (*nahiya*) of Jadida,[2] do hereby acknowledge that

1 As mentioned in the introduction, we have collected a considerable volume of local documents during our field surveys since 2005.

2 Jadida was the most populous village after Qasr in Dakhla Oasis at the time of the 1897 Population Census. It had a population of 2,410, while Qasr, Mut, and Rashda had 3,758, 1,341, and 1,191 people respectively.

 | DOI 10.1163/9789004317390_007

> each of them owes half of the total sum of two thousand, nine hundred and ten piastre (*qirsh*)[3] in governmental pure money (*'umla sagh miri*) to the honourable *Hajj* Ahmad Muhammad Abd al-Hafiz and his brothers *Hajj* Sharf al-Din,[4] *Hajj* Hasanayin, *Hajj* Khalil and *Hajj* Sulayman from Rashda, as payment for wheat and hulled rice that has been delivered to them at this price. Mansur and Miftah pledge themselves to disburse the due amount to their counterparts by the coming month as from the date on which this [pledge of debt discharge] has been issued, as attested by them and testified by the undersigned witnesses. This has been drafted on 12 Rajab 1206 coinciding with the Coptic month of Barmahat 1605.[5]

The debtors were Mansur and Miftah from Jadida, and the beneficiaries were Ahmad Muhammad Abd al-Hafiz and his brothers (Sharf al-Din, Hasanayin, Khalil and Sulayman) from Rashda. The important fact in this document that concerns us is that the five beneficiaries were notable people in Rashda and members of the future umda (village chief) family.[6] This indicates that Rashda had already been recognized as an area at the end of the 18th century.

Rashda as a Field

Next is a document with a date at the beginning of the 19th century. This is the document dated 10 Rabiʿ II 1215 (1800) concerning land lease. The document states as follows.

> The purpose of this [notarial deed], which has been duly drafted and completed by the noble Religious Council (*majlis*) at Qalamun city (*madina*) in Dakhla Oasis in the presence of our master and lord, the legal administrator (the *hakim*),[7] who shall affix his noble signature and seal above, is to certify that *Shaykh* Umran son of late Razg Allah, duly mandated by power of attorney to act on behalf of his spouse Gazza, a respectful lady of full age and capacity, daughter of the late Abd al-Salam,

3 The piastre (*qirsh*) was the currency of Egypt until 1834. It was subdivided into 40 *para*, each of 3 *akçe*.

4 *Hajj* is the honorific title of one who has performed the pilgrimage (*hajj*) to Mecca.

5 Barmahat is a month in the Coptic calendar (25 February–26 March).

6 *'Umda* in Arabic. On the post of umda, see Chapter 1.

7 On the *hakim*, see Chapter 1. In the context of the document, he probably played a role equivalent to that of *qadi* (judge).

> who gave him the authority to stand in proxy for her, as witnessed by *Shaykh* Ramadan Maʿatug and *Shaykh* Ayyub, has entered with the honourable *Shaykh* Abd al-Hafiz son of *Shaykh* Husayn Abd al-Mursil into mutual agreement in conformity with the requirements of Islamic law (*shariʿa*) and based on the free will and mutual wish of the two parties, both able bodied persons and in full possession of their faculties, whereby *Shaykh* Abd al-Hafiz will plant a plot of land located in the field (*ghuta*) of Rashda at the basin (*hawd*) of Khisas, whose four boundaries are as follows: the South is (the land of) al-Musaq; the North is (the land of) Fuqaha, the East is (the land of) Rabiʿi, and the West is (the land of) the heirs of Abd al-Sayyid *Hajj* Namr. It has been agreed that the above-mentioned *Shaykh* Abd al-Hafiz should undertake to grow various kind of plants such as *faliq al-asfar, tamr al-qaʿaqaʿ, saʿidi,*[8] and so on as he may see fit to grow in the above-mentioned plot of land, on the understanding that he would permanently be in charge of watering those trees and that he will earn in exchange for his labour twelve *qirat,*[9] that is half of the original twenty four *qirat*, while his assignor (the person who entrusted him with this task) *Shaykh* Umran, standing in proxy for his spouse Gazza, would as the owner of the above-mentioned plot of land earn the remaining twelve *qirat* (the other half). This agreement has been legally entered into and adhered to by the two parties in good faith and in conformity with the requirements of the legislation in force and the Islamic prerequisites on this date of the 10th of Rabiʿ II, 1215.

This is the document on the contract of land lease between *Shaykh* Abd al-Hafiz and *Shaykh* Umran, the proxy of his wife Gazza, who was the owner of land concerned. *Shaykh* Abd al-Hafiz earns half (12 *qirat* of total 24 *qirat*) of the produce in exchange for his labour to "undertake to plant various types of date palms and other types of plants as he may see fit to grow" on the condition that "he would permanently be in charge of watering those trees". Moreover, *Shaykh* Umran earns the remaining 12 *qirat* (the other half), as the owner of the land.

In this document, we find an expression for the fields (*ghuta*) of Rashda. The term "field" (*ghuta*) indicates that at the beginning of the 19th century, the whole area of Rashda was agricultural land, probably because of the relative abundance of water there. The field of Rashda came under the administration of Qalamun city in Dakhla Oasis (*waha*), because the document was drafted at

8 *Faliq al-asfar, tamr al-qaʿaqaʿ*, and *saʿidi* are types of date palms. On *tamr al-qaʿaqaʿ* and *saʿidi*, see Chapter 11.

9 The *qirat* (karat) is here the unit that signifies the share in the total 24 *qirat*.

"the noble Religious Council at Qalamun city in Dakhla Oasis".[10] There was supposed to be no building except cottages where peasants coming from Qalamun stayed at night in the busy farming season. They were located at the eastern side of the foot of a sand-hill (Map 9 on page 94).

Another document, which refers to the fields of Rashda, is the document dated Shaʿban 1226 (1811), which deals with the agreement on the arbitration between two parties on the water distribution of a spring (*ʿain*). The document states as follows.

In it, the first party and its partners (*shuraka*ʾ), who will be mentioned below, require 15 *amila*[11] from the total 44 *amila* of the flowing water of a spring (*ʿain*) in Dakhla Oasis Village (*nahiyat al-wahat al-dakhla*) in Qalamun, which is known as Ain al-Rahma, close to the field (*ghayt*) of Rashda.[12] The fields of Rashda contain the land enclosed within the borders in four directions; the land with one part close to Ain al-Shahid, owned by Muhammad Ismail, Yusuf Abd al-Rahman and their partners, and whose other part is close to Ain Jadid[13] owned by Muhammad Tanbur Ahmad Abdullah and their partners to the south, at the foot of the mountain to the north and east; and the land possessed by its owners to the west. However, the opponents (the second party) deny the requirement [of the first party] and appeal to take exclusive possession of the total [water of] this Ain [Ain al- Rahma].

The two parties were represented by five and eight people respectively. These 13 people were so-called notables (*mashaykh*, sing. *shaykh*) and represented the members of each party, who were called partners (*shuraka*ʾ, sing. *sharik*) to share the right (*haqq*) to use the water of the spring concerned. The members of each party are listed in the latter part of the document, and there were more than 49 and 108 people respectively.

The conflict between the two parties occurred when the first party insisted on the right (*haqq*) to use the water of 15 *amila* from a total of 44 *amila* of water flow from Ain al-Rahma. In opposition to this, the second party refused these demands and denied the first party the right to use water from Ain al-Rahma.

This was a kind of conflict over water distribution, and the reason for such conflicts was that the land was usually irrigated by more than one spring. The first party owned the land irrigated by Ain al-Shahid, which was close to the

---

10 Qalamun was one of the administrative centres of Dakhla Oasis, and the village to which Rashda belonged.

11 The *amila* is the most important concept in the division of water. It means a share of the right to use water. On this subject, see Chapters 7 and 10.

12 There are other documents, in which the fields (*ghayt*) of Rashda are mentioned.

13 For the location of Ain al-Rahma, Ain al-Shahid, and Ain Jadid, refer to Map 10–2.

fields of Rashda owned by the second party, located to the east of the office of the present Local Unit of Rashda.[14]

"The dispute between them continued for a long time without a compromise", and finally it was brought to the religious court in Asyut for arbitration.[15] As a result, an agreement between the two parties on this subject was achieved as follows.

> The possession (*wadʿ al-yad*) of the second party of the Ain concerned [Ain al-Rahma] was confirmed by the testimony of Uthman son of Abd al-Rahman from al-Amayid Bedouin, Janidi son of Miftah father of Haraq from al-Jahma Bedouin and Mabruk son of Nafd Huwaysh from al-Saadina Bedouin, who are living in Najʿ al- Jahma.[16] Then it was agreed that the Ain associated with the land defined above would be shared (*mushtarika*) by the two parties; 46 *amila* are divided and passed into their hands according to the rule, and the building and cultivation of virgin land (*mawat*) are by way of partnership (*ishtirak*). Ten and one-sixth *amila* must be given to the first party and partners (*shuraka'*) mentioned below, 28, a half and one-third *amila* to the second party and partners mentioned below, one *amila* to *Shaykh* Yusuf Muharram and *Shaykh* Hasan Umar, the *qadi*s in the village concerned, and six *amila* to the well diggers (*najjarin*) and divers (*ghawwasin*) for the restoration of the spring concerned.

This document shows the traditional way in which dispute over water distribution was resolved at the beginning of the 20th century, when it could not be arbitrated between the parties concerned themselves. It is particularly interesting on this point that the Religious Court (*al-mahkama al-sharʿiya*) had the authority to make the final decision to resolve a conflict, and the Bedouin tribespeople played an important role by giving testimony in a lawsuit in local societies in Upper Egypt and the oasis region.

In the document, the area was named the field (*ghayt*) of Rashda, and the two parties in dispute were called people of Dakhla Oasis Village (*nahiyat al-wahat al-dakhla*) at the Religious Court at Asyut. Dakhla Oasis Village signified the area administered by the present Qalamun Village.

The document also demonstrates an important fact about oasis society; social relations in oasis societies are based on the sharing of water resources by

14 On Local Units, see Chapter 1.

15 On the religious court, see Chapter 1.

16 al-Amayid, al-Jahma, and al-Saadina are the groups of settled Bedouin around Asyut.

their members. In addition, social relations based on water sharing include not only the holders of the water rights but also those who engaged in excavation, maintenance, or management of the spring concerned, such as well diggers and divers, who will be discussed in detail regarding the custom of co-operation (*ja'ala, ga'ala* in Egyptian dialect) in Chapter 7.

Moreover, this document is very interesting concerning the history of Rashda, because it shows that the members of the second party were a core group of migrants from Qalamun to the "old" Rashda who built the settlement there, as will be mentioned below. Their land was irrigated by the Ain al-Rahma, which has remained one of the wells that irrigate the land of Rashda until now (see Chapter 10).

In fact, the descendants of almost all members of the second party now live in Rashda. Its leader was *Shaykh* Muhammad Abd al-Hafiz and his son, named *Shaykh* Ahmad *Hajj* Muhammad Abd al-Hafiz, who was the first umda (village chief) of Rashda after its independence from Qalamun, as will explained below.

According to the informant, the agricultural land belonging to Rashda was owned by the three largest families living in the village of Qalamun in those days. One of these was the family of *Shaykh* Muhammad Abd al-Hafiz mentioned above. The land was cultivated by peasants who lived in Qalamun, and they went to and from the fields in Rashda every day.

The document, dated 23 Sha'ban, 1268 (1852), sheds light on the situation of the fields of Rashda in the first half of the 19th century (Figure 1). This is a *waqf* (religious endowment) document. The document is written in the proper form for the *waqfiya* (*waqf* document) at the religious court (*mahkama*) of Qalamun whose *qadi* was Abu Bakr Ayyub.

The donor (*waqif*) of the *waqf* and the writer of the *waqf* document was Jamal al-Din, villager of Qalamun. The third-generation descendants of his family now live in Rashda. Jamal al-Din, son of the honourable Shadhli Jamal al-Din from Qalamun, drafted this document according to Islamic law; 35 *waqf* materials (*mawquf*) are listed. It details five buildings, 22 pieces of land, and eight moveable items of property.

The *waqf* property holdings are listed according to location, shown according to the surrounding places or buildings, as they are in the *waqf* document. Therefore, we can identify their approximate locations on the present-day map.

Four of the five buildings, with the remaining one located in a quarter (*hara*) in Qalamun, are buildings related to present-day Rashda. Two of the four were located in present-day "old" Rashda, and the other two in the fields (*ghuta*) of present-day Awina. Awina belonged to Rashda in those days.

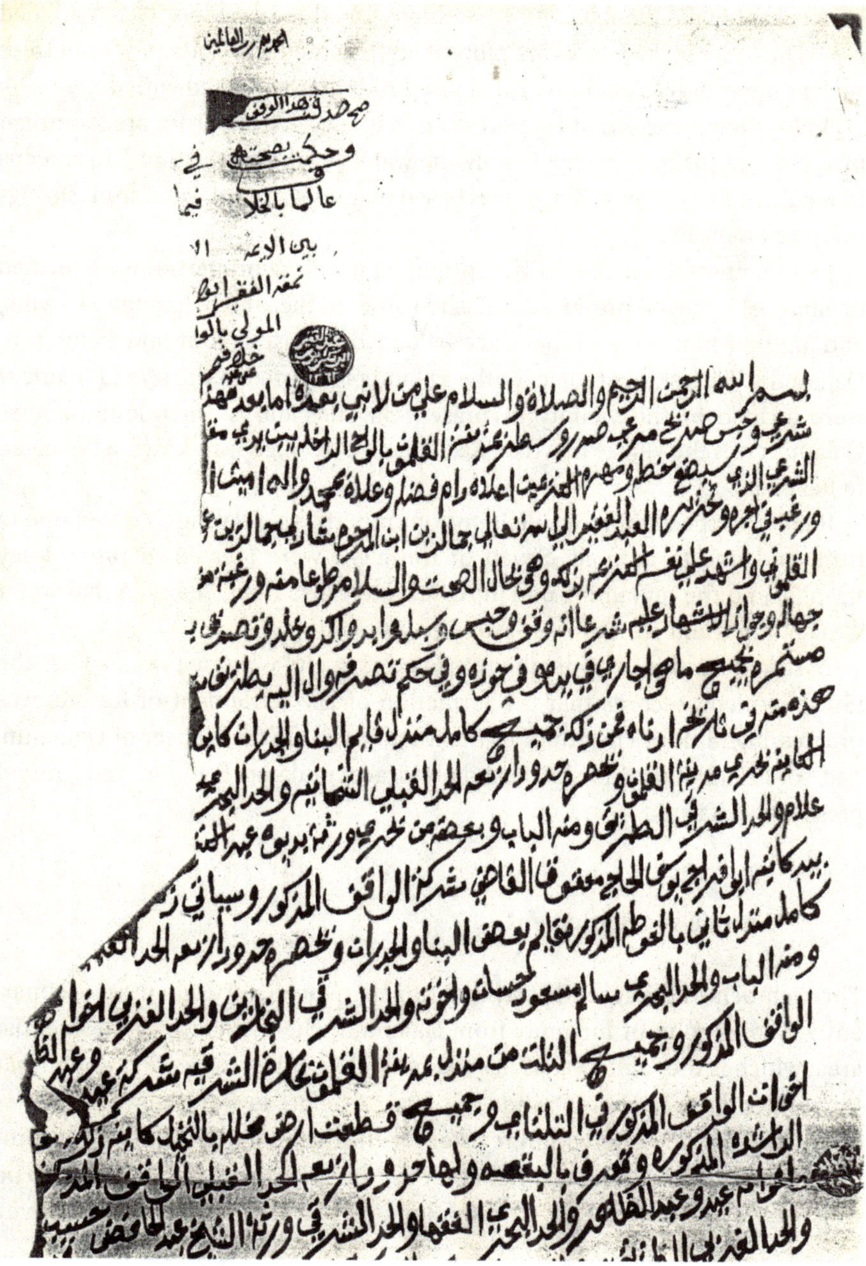

FIGURE 1 *The beginning of the waqf document dated 23 Sha'ban, 1268 (1852)*

The main *waqf* property holdings are 22 plots of land, all of which are agricultural, and their uses are briefly explained for each plot. Four of the properties are "pieces of land planted with something", and three are plots for unspecified uses. The other 15 properties are plots of land planted with dates and fruit trees such as olive and lemon trees. All 15 properties are associated with dates.

As for their locations, it is interesting that 14 of 22 properties are identified by the names of springs that supply the water for their irrigation. This means that cultivation in oasis villages was based on underground water from springs (Map 22 on page 195).

Judging from the names of the springs, 14 of the 22 properties were located on land belonging to present-day Rashda, four on the land belonging to Awina, and another four on land that once belonged to Rashda, but now belongs to Qalamun.[17] This shows that all of the 22 land plots listed in the *waqf* document were located in the vicinity of present-day Rashda. As mentioned above, Qalamun was the village to which Rashda once belonged, and Awina belonged to Rashda.

Finally, of the eight movable items of property, six relating to water, one to fruit, and one to animals. Seven of the eight were located in present-day Rashda, and the remaining one in Awina. Here also it may be seen that water was an important item of *waqf* property.

In summary, this *waqf* document, dated 1852 and written 40 years after the 1811 document, records that the formation of the settlement of Rashda was progressing rapidly. The donor of the *waqf*, Jamal al-Din, villager of Qalamun, had some buildings in "old" Rashda and accumulated lands in and around present-day Rashda.

Rashda as a Settlement

The members of the second party mentioned in the 1811 document were apparently a core group of migrants from Qalamun who settled in the residential area, which is the "old" area of Rashda today. Moreover, there is no doubt that the residential area of old Rashda was formed before 1278 (1861/62).

In that year, the population register for the settlement of Rashda was produced; this will be discussed in Chapter 8. From that time, Rashda began to be called a *nazla* (settlement). The implementation of the population survey

17 The springs that provided water to the cultivated lands belonging to Rashda in the document were Ain al-Balad Mustajadd, Ain al-Balad, Ain al-Jumayza, Ain al-Shishlana, Ain al-Rahma, and Ain al-Musid.

shows that Rashda was substantially a "village", although for administrative purposes, it still belonged to Qalamun. In fact, conflicts and judicial cases were judged in the Religious Court (*mahkama*) in Qalamun.

As mentioned above, the agricultural land belonging to Rashda was owned by large families living in the village of Qalamun in the first half of the 19th century. Migrants from Qalamun settled in Rashda, at first as peasants who cultivated the agricultural land possessed by the three largest families. They were given the land for their livelihood by landowners in return for their agricultural labour.[18]

In 1862, 10 years after the writing of the *waqf* document discussed above, an arson attack occurred in Rashda. A document was drafted on this arson on 28 Jumada II 1279 (1862)(Figure 2). Below is the translation of this document.

> On 28 Jumada II 1279 [December 21, 1862], after a fire had broken out in the settlement (*nazla*) of Rashda (...) and the inhabitants of the settlement had been unable to identify the perpetrator of the arson or any suspect (....)., *Shaykh* Ahmad Abd al-Hafiz, *shaykh al-balad* (village shaykh) and other notables of Rashda settlement decided to compile a list of the inhabitants of each quarter (*darb*) in the settlement and to summon them to their presence for testimony, in order to take a pledge from them. There was no objection from anyone about this, and the inhabitants of each quarter of the Rashda settlement were duly identified as follows [in Table 14].

When [the notables of the settlement] assembled on the date mentioned below, they acknowledged that they had not known who the perpetrator of the arson was and made a pledge that the inhabitants of each quarter would henceforth keep a close watch, and that if, God forbid, incidents such as arson attacks should occur in some part of the said quarters of the settlement or the neighbouring areas, they would take it upon themselves to disclose the name of the perpetrator and would assume liability for failing to abide by this commitment. This pledge was taken in the presence of the signatories of an affidavit signed to this effect on 28 Jumada II 1279. Then the local people adhered to this affidavit, pledging that they would henceforth be responsible for protecting houses and other constructions during night-time or any other time, and that

18 Refer to the case of the land transactions in Bir Nakhil District in Chapter 12. Moreover, the peasants who had migrated from outside and settled in the "old" area and the Nazla area bought the land, which they cultivated, from the landowners.

TABLE 14 *Inhabitants of Rashda by* darb (*quarter*)

Name of darb	Number of households (*usra*)	Number of persons
Darb Shaykh Ahmad Hajj Muhammad and his uncle's sons (*awlad*)	5	9
Darb Razaziqa	6	13
Darb Naji····	1	1
Darb Shadhli Abd al-Bari and his uncle's sons (*awlad*)	1	3
Darb Abd al-Mahdi and his group (*jama'a*)	5	11
Darb Abd al-Wali Saghir	3	8
Darb *Hajj* Ahmad Abd al-Hafiz	8	11
Darb Abd al-Maqsud Rifai	3	3
Darb Salih Sayyid and Abd al-Nafa and Asham Allah and Abd al-Sattar	5	8
Darb faced northernly to the house (bait) of Marzuq al-Hijji	11	23
Darb faced southernly to the house (bait) of Marzuq al-Hijji	6	8
Darb Khayr Allah and Atiya	2	5
Darb Timsah	3	3
Darb Salim Masud	3	5
Darb Najjarin	8	12
Darb Saad Allah Said (the family of Ababida) *Hajj* Hasan	7	14
Darb Awad	5	5
Darb Ibrahim al-Mallakh	1	1
Darb Nahiyat al-Muzayyinin	2	5
Darb Abd al-Nasif	1	2
Darb Jahmi Abada and his group (*jama'a*)	6	9
Total 21 darbs	92	159

Note: It is not clear why the total number of persons in the table is different from that calculated in the document. As for the location of *darb*, see Map 19 on page 172.

if any mishap should occur, they would abide by arbitration decisions and provide compensation as the case may require.

The table of inhabitants by quarter (*darb*), which is attached to the document, is surely supposed to be drafted, based on the population register of 1861 for Rashda,[19] which is examined in Chapter 8. According to the table, 92 households (*usar*, sing. *usra*) including 159 adult men resided there.[20]

19 This document was dated 9 Jumada II 1278 AH (1861 AD).

20 *Fard* (plural *afrad*) in Arabic. Judging from their number, they appear to be adult men who may have been responsible for the security of the settlement of Rashda.

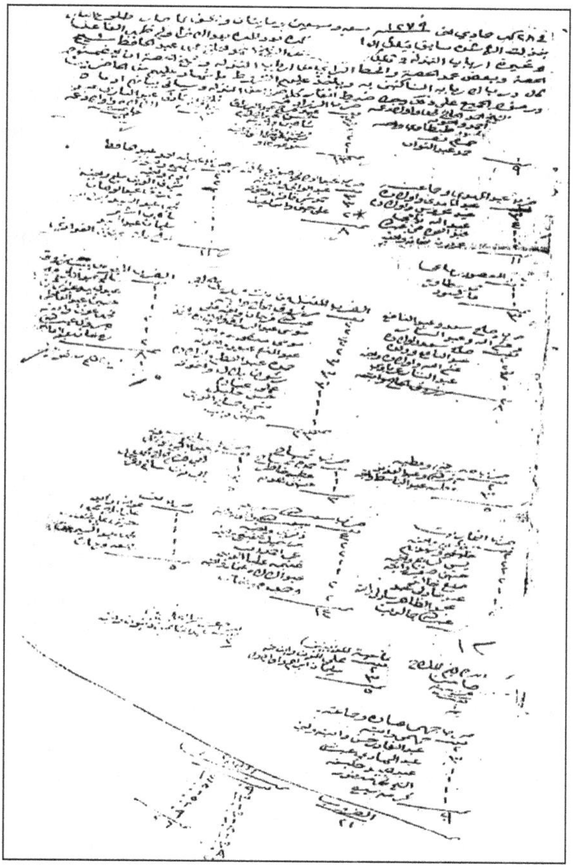

FIGURE 2
Document dated 28 Jumada II 1279 (1862) on arson

The document explicitly shows that Rashda already had the substantial structure of a village and was recognized as a village until 1861, although it was not still formally recognized as an administrative unit.

In the document, the settlement (*nazla*) where the act of arson occurred was also named the quarter (*hissa*) of *Shaykh* Ahmad *Hajj* Muhammad Abd al-Hafiz who resided in Qalamun at that time. He was *shaykh al-balad* (village shaykh) at that time, and since then, his family has been called the "family (*ʿa'ila*) of *shaykh*".

Then, after the independence of Rashda from Qalamun, the first umda (village chief) of Rashda was his son, *Hajj* Musa Ahmad Muhammad, son of *Shaykh* Ahmad.[21] The document confirms that the settlement (*nazla*) of

21 After Musa Ahmad Muhammad, the successive post of the umda in Rashda has been held by the members of his family until now: Abu Bakr Khalil son of Muhammad Abd al-Hafiz

Rashda was recognized as one of the quarters that belonged to the village of Qalamun.

The settlement (*nazla*) was composed of 21 quarters (*darb*). The inhabitants of the settlement seemed to form a kind of community that was responsible for its own security. As mentioned above, the core of this community was formed by migrants from Qalamun.

Migration from outside was continuing. Some peasants who migrated from the village of Balat (one of the oldest villages in Dakhla Oasis) settled in the area below the sand-hill, which is now called Nazla, on the northern side of the "new" area of Rashda. The Nazla area was a marshland at that time. They also came there to seek land and water for their livelihood.

Landscape of "Old" Rashda

The "old" Rashda was on the sand-hill west of the "new" Rashda, which is the present Rashda. Almost all houses in the "old" Rashda are now abandoned and are largely in ruins. However, clear traces of the former age remain.

The "old" Rashda was famous for the gardens of dates and fruits that surrounded the sand-hill, on which the "old" Rashda was located at the beginning of the 20th century.[22] However, according to the descriptions included in the documents that are referred to above, the landscape of "old" Rashda had already been formed in the middle of the 19th century.

The settlement itself of the "old" Rashda on the sand-hill was a complex of quarters (*darb*), as shown in the table mentioned above, where the information on inhabitants is arranged by quarter. Fortunately, it is possible for us to visualize the location of the quarters by linking the historical information collected during the field surveys with digital maps.

For this purpose, with the support of the informant, we attempted to locate each quarter on the digital map, asking the village notables and the villagers who had lived in the "old" Rashda or who still owned houses there. Then we checked the owners of houses and produced a list of those in the "old" Rashda. Map 8 shows the location of quarters.

---

cousin of *Shaykh* Ahmad, Husayn Khalil brother of Abu Bakr, Hasan Sayyid Hasanayn Muhammad Abd al-Hafiz from the family of Shaykh Ahmad, Muhammad Imam Khalil from the family of Shaykh Ahmad, and the present umda, Kamal Muhammad Imam.

22 Refer to the report on villages in the Libyan Desert by Godfrey Sykes [Sykes 1912: 732] in Chapter 6.

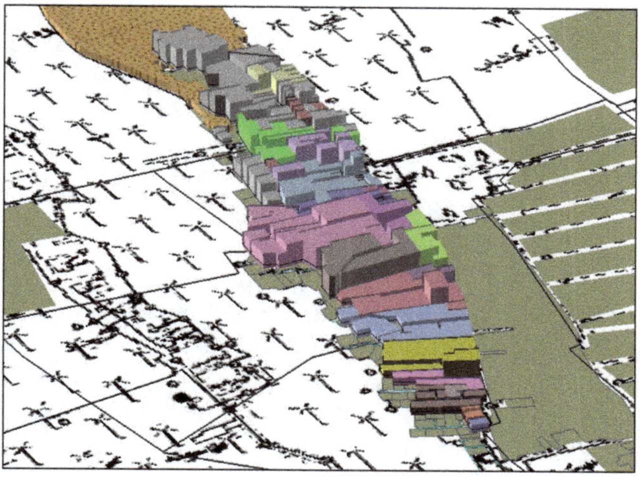

MAP 8
Complex of quarters (darb) *in the "old" Rashda.*
Source: Based on the field surveys conducted in 2008–2009.

It can be confirmed from this map that the "old" Rashda, although it had no walls, had a similar shape to traditional Egyptian villages, divided into 21 quarters for the residence of inhabitants in 1862. According to interviews with the elders, the pattern of habitation in the "old" Rashda remained basically the same until the villagers moved to the "new" Rashda.

The quarter was named *darb*. In its strict meaning, a *darb* is a blind alley that is also called a *hara*, *sikka*, or *zuqaq*, in comparison with *shariʿ*, which means the open main street. However, it also signifies a common place that people living in the blind alley share as a living space. Therefore, this space is sometimes translated as "neighbourhood".

In this meaning, the *darb* is a unit of residential space with an entrance door that was closed at night for security. It was generally inhabited by members of one or more families (*ʿaʾila*) and was named after the chief of the largest family; for example, the Darb *Hajj* Ahmad was named after the head of the future umda family.

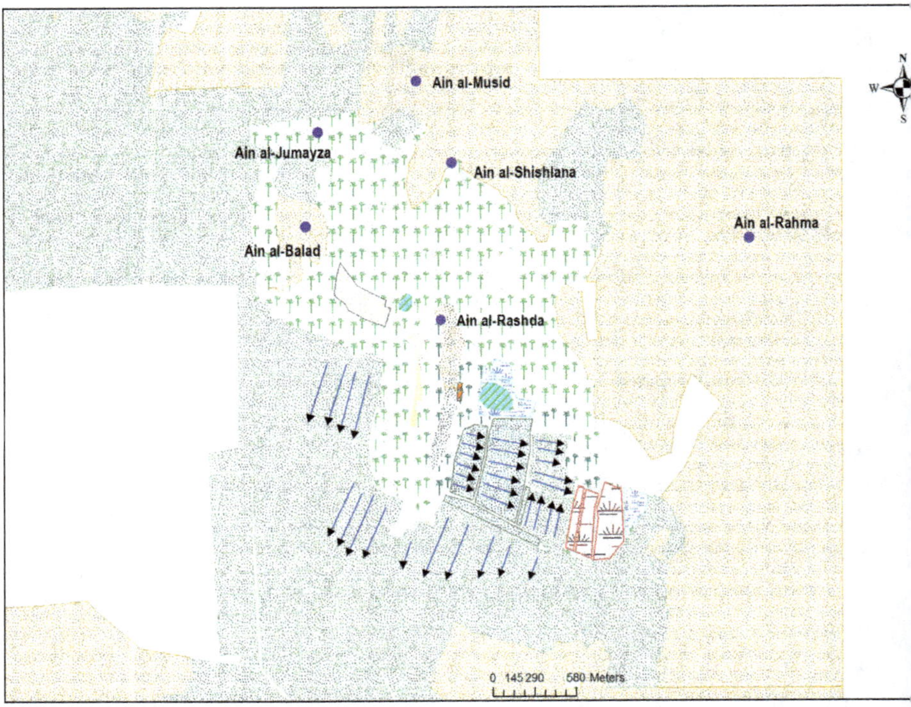

MAP 9 *Rashda at the beginning of the 19th century. Source*: Based on the information collected by informant.

PHOTO 3 *The place where the farmers of Qalamun built their cottages to stay during the agricultural season*

PHOTO 4 *View of the "old" Rashda on the sand hill (Remains of the residence of Khidr Family on the left)*

PHOTO 5 *Remains of the residence of the* Sheykh al-Balad *Khidr Hassinin who was a brother of the first Omda in the "old" Rashda*

PHOTO 6
A darb *in the "old" Rashda. This shows the entrance of Darb al-Najjarin and the former two-storey house of the Jamal al-Din family.*

PHOTO 7 *Remains of Ain Barqas near the remains of Ain Rashda (Qara District)*

PHOTO 8 *Remains of Ain Jamayza*

CHAPTER 6

Development of Rashda

Rashda was recognized administratively as an independent village from Qalamun Village at the end of the 19th century. Since then, Rashda has developed as an agricultural village and this is reflected in the landscape of "New Rashda", which was formed in the agricultural fields as a new settlement area of the village under "Old Rashda" on the sand-hill. This chapter deals with the process of Rashda's independence from Qalamun, its expansion and the formation of "New Rashda", and the change of residential patterns and village lifestyle, mainly using the statistical data on households and the geographical information collected during field surveys.

Independence of Rashda from Qalamun

In 1897, a population census was conducted. It was the first substantial modern census conducted in Egypt.[1] Rashda appears in the population census of 1897 as Nazla (Settlement) Rashda, separate from Qalamun village (*nahiya*) in the Mut district (*ma'muriya*), Asyut province (*mudiriya*).

Nazla Rashda had a population of 1,191 people (576 men and 615 women), of whom Nazla Rashda itself had 936 and the associated settlement (*'izba*) of Awina Shaykh Abdullah had 255. Qalamun had a population of 1,704 (759 men and 945 women), of whom 1,447 lived in Qalamun itself, and 153 and 104 lived in the two associated settlements of Ain al-Qada and Ain al-Roda, respectively (1897 Population Census: 274). The fact that Rashda appears as a separate entity means that Rashda was recognized as a village in 1897,[2] although for administrative and judicial purposes, it belonged to Qalamun. In fact, the inhabitants were ruled by Shaykh al-Balad, and the conflicts and judicial cases were judged in the Religious Court (*mahkama*) in Qalamun.

Judging from the size of the population in 1897, there is no doubt that Rashda already had the landscape of a village and was recognized as such

1 The first modern population census was conducted in 1882. However, this census was a form of experiment, and the 1897 Population Census is considered to have been the first substantial population census in modern Egypt.

2 At that time, 221 buildings were surveyed.

 | DOI 10.1163/9789004317390_008

when the population survey was conducted in 1861,[3] which is discussed in Chapter 8. The settlement of "old" Rashda had the structure of a complex of quarters like a traditional Egyptian village, as explained in Chapter 4.

It is not clear when Rashda came to be called Rashda Village (*qarya*) and was registered as an independent village. In any case, immediately after the population census in 1897, Rashda was apparently formally separated from Qalamun, and it became an independent village in 1899. In *al-Qamus al-Jughrafi* by Muhammad Ramzi, Rashda is mentioned as a new village that had been a settlement attached to Qalamun that was separated to become an independent village in that year (Ramzi 1968: part 2, vol. 4, 243).

However, in a kind of administration manual published in 1904, Rashda is called the "settlement" of Rashda (*nazla al-rashda*) with a population of 1,191,[4] and no information such as the number of date palms, volume of water, or number of springs is provided. In contrast, it reports that Qalamun, which is called the "village" of Qalamun (*nahiya al-qalamun*), had 26,378 date palm trees, 154 *qirat* of water (1 *qirat* is 100 cubic metres), 50 springs, and 2,704 inhabitants (Hunayn 1904: 210).[5]

In 1907, eight years after Rashda became a separate village, there were 223 houses and 1,223 inhabitants (604 men and 619 women) according to the population census of that year.[6]

A Village of Migrants

The inhabitants in those days were migrants from elsewhere. According to the household survey of 2005, one-third of the households in present-day Rashda are originally from outside the village. They came mostly from Balat, Mut, or

---

3 In a document dated 28 Jumada II 1279 (1862), 92 households (*usar*, sing. *usra*) are mentioned. This means that about 500 people inhabited the "old" Rashda, indicating that the average household size was approximately five people.

4 Judging from the size of the population (1,191), the demographic data to which Hunayn refers are based on the 1897 Population Census.

5 According to this information (Hunayn 1904: 210), there were 12 villages in Dakhla Oasis in 1897: Qasr (with a population of 3,759), Budkhulu (583), Mushiya (927), Jadid (2,392), Qalamun (1,704), Nazla al-Rashda (1,191), Hindawi (643), Mut (1,341), Maasara (1,078), Asmant (1,037), Balat (787), and Nazla al-Tenida (843).

6 Nizarat al-Maliya (1909) provides data on the basic indicators of the social economy by village, which were collected by the survey that accompanied the 1907 Population Census. In it, Rashda is described as a village with one *'izba* (settlement) and 13 springs and wells, 23,114 taxed palm trees, 382 Egyptian Pounds of taxes on land and palm trees, and 590 tax payers. It is instructive that there are no data on cultivated land, such as area or tax paid on it, because water is a more important factor than land in oasis villages.

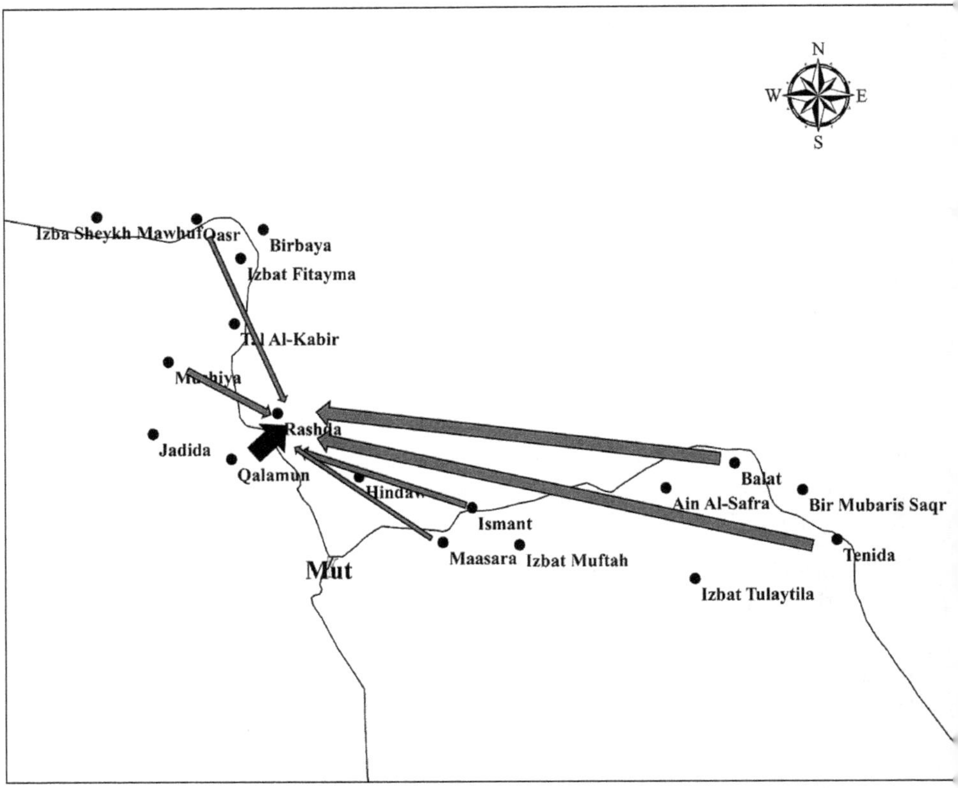

MAP 10 *Villages of origin of the households in Rashda. Source*: Information from the informants.

the neighbouring villages in the Dakhla Oasis region such as Awina, Messaud, Jadid, and Qasr, along with some households from provinces such as Asyut in Upper Egypt (Map 10). It may be noted that there are families (*ʿaʾila*) classified as originally from Rashda but known to have come from outside the village[7].

The family of J shows the typical pattern of migration to Rashda by migrants in those days. J's family settled in Rashda as agricultural workers for the Nazla area, as did those from Balat.

J (aged around 50 in 2005) and her husband lived in the Nazla area with her husband's mother Z, who was born in 1918. Z was born in Rashda, but her father was born into, and was originally from, one of the settlements (*ʿizba*) belonging to Balat village (Izba al-Safra).

7 For example, the family of AB was originally from a village in Minya called Bani Abid.

Her father came to Rashda 100 years ago, around 1910–1915. At that time, Sanusi troops were stationed at Dakhla Oasis, and the soldiers were forcing the villagers to provide food and agricultural produce.[8]

J, who is also originally from Balat, remembers that her grandfather was killed by one of the soldiers. According to Z, her father came to Rashda to escape the violence of the Sanusi troops and because there was plenty of water and a demand for labourers in Rashda.

Her father settled in one of the cottages provided by the umda of Rashda in Nazla and worked on the umda's farm. He was given food, some agricultural produce, and a house for cultivating and taking care of the gardens. After some years of working on an unpaid basis, her father became a waged labourer on the same umda's farm.

Z married one of the villagers, who was also originally from Balat. After the revolution around 1960, Z and her husband and their sons bought the land sold by the government after the construction of the *bir* (well), and became self-employed farmers. In 1962, they also bought the land in Nazla from the umda and constructed their own house.

Farmers in the Settlement of Rashda

Most of the inhabitants in the second half of the 19th century are thought to have been poor people, just like the migrants in the first stage of the settlement mentioned above. According to interviews with the village notables in Rashda and Qalamun, their ancestors were "la tamlik wa la yaamal" (landless and unemployed). They obtained the land in Rashda either from the family of the umda or the landowners in Qalamun (there were lands owned by the villagers of Qalamun) either directly or after working as share-croppers.

The document dated Safar 1270 (1853/54) shows the circumstances in which the inhabitants in those days lived. The document consists of three tables written for the landowner to show how the land was managed. The purpose of the tables was to confirm the shares of the rice crop allocated to the landowner and farmers (fellahin) working on his land.

In this document, the yield from the rice crop was calculated by well. The rice crops harvested from the land of the landowner were brought to the jurn (threshing place) named after the landowner. For example, the rice crops of Ahmad, who was the head of the future umda family, harvested from land irrigated by the water of Ain al-Jumayza, were brought to the jurn named "Jurn al-Ahmad". The rice crops, which were brought to the Jurn al-Ahmad, were divided into three shares, and each was measured.

8 On Sanusi troops, see Excursus 1 in Chapter 1.

We do not know why the yield was measured in three lots, whether they were harvested from three different plots of land possessed by Ahmad, or whether they were brought to the jurn at three different times. However, the latter scenario appears to have been the case, because we find some farmers of the same name present on all three occasions. In any case, this part of the document is concerned with the land irrigated by the water of Ain al-Jumayza owned by Ahmad.

In Rashda, the jurn was generally meant to indicate the open land that was occupied through possession (wadʿ al-yad) and was equipped with a store (shuna) for threshing and temporary storage of crops after the harvest, as shown in Photo 9. However, in the case of Jurn al-Ahmad in the document, which means the Jurn of Ahmad, it was not land designated for the threshing and storage of crops, but a part of the vacant cultivated land after harvest, near the land possessed by Ahmad, as shown in Photo 10. The rice crops were brought here because unlike wheat, they were too heavy to transport.

As mentioned above, the yield from the rice crop was divided into three. The total volume of Jurn al-Ahmad was 146 *haml*.[9] The volume of the three shares was 76 *haml*, 40 *haml* and 30 *haml*. The net share of the landowner's family (called *bait*[10]), excluding six *haml* of the reserve (*dhariʿa*) for seeds [for the following year] and liquidation, was 78 *haml* (40, 20, and 18 *haml*, respectively, for each of three shares) or about 56% (57, 50, and 60%, respectively, for each of three shares) in total.

The remaining crops were the shares of farmers (*fellahin*) working on the land of the landowner Ahmad. Of the 76 *haml*, 30 *haml* were divided among six farmers (6, 3, 3, 6, 3, and 9 *haml* respectively), Of the 40 *haml*, 20 *haml* were divided among six farmers (4, 2, 2, 4, 2, and 6 *haml* respectively), and of the 30 *haml*, 12 *haml* were divided among seven farmers (2, 1, 1, 2, 1, 3, and 1 *haml* respectively).

The farmers were supposed to work the landowner's land and to obtain rewards in kind in return for their labour. In addition, the reserves of seeds [for next year] and liquidation were deducted from the share of the landowner. In this sense, they appear to have been share-croppers (*muzariʿin*).

A *muzaraʿa* contract is usually taken to be a share-cropping contract between a landowner and a tenant farmer. Landowners have the right to conclude such a contract, because the land is the essential factor in the modern landownership system.

9 1 *haml* = 4 *wiba*; 1 *wiba* = 10 *misha*.

10 This literally means "house". See Chapters 3 and 9.

PHOTO 9 *Entrance to a* jurn *in Ain al-Balad (Sanquha) District*

PHOTO 10 *Open field after the rice harvest in the Ain Rashda District (right) and Ain Rahma District (left)*

However, before the establishment of private landownership at the end of the 19th century in rural Egypt, *muzaraʿa* were not share-cropping contracts in the modern sense but were considered to be contracts for cultivation among farmers, who each brought some production resources such as land, labour, water, animals, and seeds, and shared the agricultural products.[11]

At a glance, the method of cultivation described in the document is cultivation based on a share-cropping contract. It is certain that the shares of revenues of rice crops harvested in the fields were divided between the landowner and the farmers working on his land. However, the contract was not drafted in the style of the book for the management of a farm.

The share of each farmer was individually calculated and recorded by his name, unlike the account of landowner. In this sense, the farmer in the document could not be identified as a "share-cropper", who would be a party in a modern share-cropping contract in relation to another party, the landowner.

In any case, at the beginning of the 20th century, Rashda was registered as one of the formal villages (Map 14).

Rashda in the First Half of the 20th Century

In the first half of the 20th century, Rashda was famous for having many palm trees and much fruit. In a report by a traveller to Dakhla Oasis, we find the following description of Rashda in 1911/12 (Sykes 1912: 732).

> Our first stop was at Rashda, and we found this to be by far the most attractive and prosperous looking of all the villages we saw. It stands, as is common, upon a hill, and is surrounded by very fertile looking palm groves and walled gardens, in which oranges, lemons, bananas and other fruit abound. The village itself contains two small flour-mills, worked by oxen, a mosque, a school, and a number of neat-looking houses. We were taken to the flat roof of one of these houses by a head-man of the place and got a splendid bird's-eye view of the groves, wells and cultivated land.

A similar description is found in an inspection report on drinking water in Egypt published in 1931 (Azadiyan 1931: 69).

> This village (Rashda) adjoins Budkhulu village to the north, and Qalamun village and Hindawi village lie to the south-west and east. Rashda village is famous for its rich gardens, and trees bearing oranges, lemons,

11 On this subject, refer to the discussion based on a series of land laws published in the middle of the 19th Century (Kato 1993: 175–181, 193–198, 410–412).

PHOTO 11
Photo entitled "Rashda from a hilltop." Source: Azadiyan 1931:732.

PHOTO 12 *View of Rashda from the hilltop today*

> pomegranates, apricots (*mishmash*), mulberries (*tut*), guavas (*gawafa*), and other fruit. Many have been planted, and there are many date palms. The fertility of its land is undoubtedly caused by the presence of much water, and we confirmed during our visit to the village that all the wells (*abar*, sing. *bir*) had a lot of water. Some of the many wells in Rashda are "old", and some have been "newly" dug. The most famous of these is the well (*bir*) named Bir Majnun.

However, the demographic trend shows the opposite situation in Rashda in those days. It seems to show that in spite of such descriptions of Rashda's riches, the first half of the 20th century was a period of social and economic stagnation, at least from the demographic viewpoint (Vivian 2008: 125, Fakhry 1974: 86).

This can be confirmed by Table 15, which shows the demographic trends of villages in the district (*markaz*) of Dakhla from 1882 to 1996. In fact, as shown in Table 16, 93% of the household heads reported that they came from Rashda. Every village notable who was interviewed during our field surveys in Rashda said that there was no period of rapid migration from outside the village after the formation of the independent village.[12]

The stagnancy of demographic trends was supposedly caused by the uncertainty of the water supply. As discussed in detail below, five springs in Rashda dried up in 1937, and the umda had to petition the authorities for a new well to be dug.

However, on the other hand, the wave of the modernization was certainly coming to villages in Dakhla Oasis. Electric well pumps were introduced to Rashda in 1941, and motor cars first appeared in the Dakhla region in 1925.[13] Previously, transportation over long distances had relied on camels, and it took 16–20 days to travel from Rashda to Cairo.[14]

Formation of the "New" Rashda

The turning point in the development of Rashda was the 1960s, when the New Valley Project began. Things began to change. The houses in the "old" Rashda, which were built of mud, were decaying, and the population there was thought to exceed its capacity. As a result, the inhabitants gradually built new houses outside the "old" area and moved into them from the 1960s.

In fact, the number of houses in present day Rashda—namely the "new" Rashda—rapidly increased in the late 1960s, as is shown in the number of residential buildings by year of construction (Table 17).

After the shift in residential space from the "old" to the "new" Rashda in the latter part of the 1960s, migration from outside Rashda became insignificant,

12 According to the demographic data on the number of males and females in the population and their birth places in the 1907 population census, there were more females than males in the population. This may suggest that Dakhla Oasis was already a region that people tended to leave in the early 20th Century.

13 See Excursus 1 in Chapter 1.

14 On transportation in earlier periods, see the travel guide by Wilkinson (1847). It contains a short reference to Qalamun (Wilkinson, 1847: 247).

TABLE 15 *Demographic trends of the main villages in Dakhla Oasis from 1882 to 2006*

	1882	1897	1907	1917	1927	1937	1947	1966	1976	1986	1996	2006
Mut	1,213	1,341	1,522	1,514	1,678	2,094	2,513	4,792	8,122	11,890	16,252	20,293
Ismant	676	1,037	1,137	1,201	1,207	1,238	1,277	1,927	2,169	2,737	2,690	2,989
Jadida	2,137	2,410	2,536	2,473	2,228	2,631	2,789	2,781	2,991	3,369	3,142	3,662
Rashda		1,191	1,223	1,337	1,319	1,543	1,739	3,306	4,398	5,574	4,364	5,423
Qasr	3,539	3,758	3,602	3,241	2,803	3,246	2,155	3,321	5,813	5,257	3,723	10,122
Qalamun	2,788	1,704	1,994	1,853	1,679	1,756	1,587	2,213	2,272	2,184	2,128	1,812
Maasara	926	1,078	1,060	1,029	1,084	1,212	1,202	2,065	2,374	3,031	3,282	3,911
Mushiya	780	818	953	965	966	959	1,146	1,233	1,571	1,778	2,146	2,495
Hindawi	500	643	764	847	790	953	1,038	1,659	2,174	2,774	3,062	3,942
Budkhulu	572	583	456	322	332	410	537	861	1,113	1,216	1,523	1,719
Balat	2,162	1,784	2,219	2,203	2,102	2,564	2,668	4,064	4,952	5,492	5,126	6,397

TABLE 16 *Which generation of your family first lived in this place (village)? (household head) (2005)*

	%
Your generation	0.9
Parents' generation	0.4
Grandparents' generation	0.4
Before grandparents' generation	5.3
Not moved	93.1
Total	100.0
(Number)	550

Source: Household Survey 2005.

TABLE 17 *Number of occupied residential buildings by year of construction (unit: buildings) (2006)*

	Number
1955–59	4
1960–64	11
1965–69	132
1970–74	82
1975–79	101
1980–84	131
1985–89	103
1990–94	109
1995–99	192
2000–04	105
Total	970

Note: Year of construction refers to the ground floor when the buildings have more than one floor.
Source: Additional Sampling Survey 2006 for the information on the construction year.

because as mentioned above, almost all the heads of household surveyed were born in Rashda.

The change was accelerated by a state policy, called "developmentalism", which was promoted under the socialist regime of President Gamal Abdel Nasser. In 1959, a construction campaign began, and many wells (*bir*) for cultivation were subsequently constructed by governmental organizations or foreign companies.

This policy is reflected in the construction boom in the "new" Rashda. There were a total of 970 buildings for residential use at the time of the survey in 2006. There are only 15 residences that were built before 1964. Construction of most residences began after 1965 and has continued uninterrupted until the present day. Most of the newly built residences are located in the eastern area, where vacant land is found.

This demonstrates that the residential space of Rashda has shifted from the "old" to the "new" Rashda. The state policy on local administration that established the Local Unit of Rashda in 1970 apparently encouraged this shift.

In other words, this shift was caused by the combination of the demolition of old mud brick houses and the increased population on the one hand, and the restructure of rural administration by the state based on "developmentalism" on the other.

Expansion of the Village Space

The residential area of Rashda has continued to develop until now. We can summarize its history according to the following four periods, referring to the description in Chapters 5 and 6. Map 11 is a conceptual map of the residential development of the village of Rashda from the time of its formation until today. Map 12 shows the residential space of Rashda at the time of our field survey in 2005.

1. Until the beginning of the 19th century, what is now the residential area was agricultural land belonging to Qalamun village, and there were no buildings except for the cottages at the foot of a sand-hill where the peasants of Qalamun spent the night during the agricultural season.
2. In the first half of the 19th century, the settlement area on the sand-hill was created, and peasants from Qalamun village migrated to it. They were joined by migrant agricultural labourers from Balat village (one of the oldest villages in Dakhla Oasis) in the Nazla area in the second half of the 19th century. The Nazla area was a marshland at that time.[15]
3. Construction of houses by the inhabitants of the settlement ("old" Rashda) by the sand-hill began on the agricultural land on the western side (the present Qara area), around 1940, and on the eastern side of the sand-hill ("new" Rashda), in the 1960s. Construction in both areas was undertaken because of migration.

---

15 The "new" area was agricultural land owned by the umda and other families. According to the informants, the plots there were sold by the umda and other families because water flowing from the old spring (*'ain rumani*) that supplied this area became slower. In addition, new wells (*bir*) were constructed, and new land became available in other areas. There were two springs (*'ain*) that provided water: the old spring, which provided water to plots in the "new" area, dried up around 1970, and another spring provided water to the plots of Izba Shamrukh.

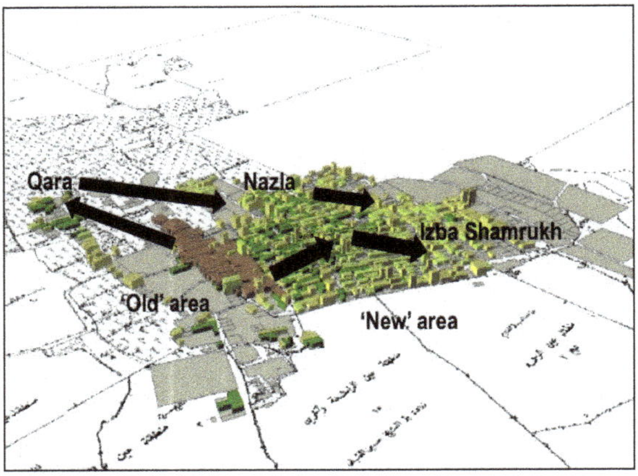

MAP 11 *Conceptual map of residential development in Rashda (19th century to the beginning of the 21st century). Source*: Kato and Iwasaki 2008: 50.

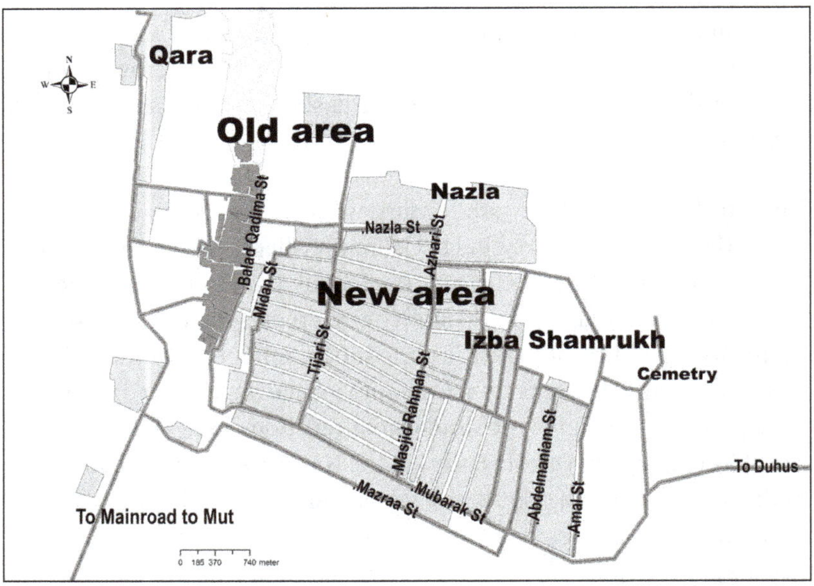

MAP 12 *Residential space of Rashda (2008). Source*: Kato and Iwasaki 2008: 25.

4. From the 1980s, more houses were constructed in the east of Izba Shamrukh and in the eastern part of Nazla.[16] The inhabitants of "new" Rashda and Izba Shamrukh migrated to these two areas.

---

16 Construction of the houses first took place on the plots beside the "old" area, horizontally alongside the plots from west to east. The plots became the residential area, and the dike (*jisr*) became the street. Then the construction shifted to Izba Shamrukh. The plots at the northern part of Izba Shamrukh were sold first, because they were at the end of a water canal.

Pattern of Residential Ownership

According to population censuses, the independent village of Rashda did not report a rapid increase in population from the 1970s. In 1976, the number of inhabitants of Rashda was 4,398. In 1996 and 2006, the total population was 4,364 and 5,423, respectively.

However, the village has expanded its residential space. Almost all residences built after the 1980s are situated in the Izba Shamrukh area (see Map 17 on page 148). This coincides with the pattern of residential ownership. Of the 83.3% of householders who own their residences, 69% declared that they purchased it (Table 18). This is different from the common pattern in rural areas, where acquisition through inheritance from parents is most common.

TABLE 18 *Ownership and method of acquisition of residences (household) (2005)*

		%
Ownership	Owned (totally paid)	83.3
	Owned (partially paid)	2.2
	Rent	3.1
	Rent (furnished)	0.0
	Gift	9.3
	Other	2.2
	Total	100.0
	(Number)	550
Acquision	From family/relative	69.0
	From friend/acquaintance	14.9
	Through samsar (real estate agent)	0.4
	Inheritance from parent	14.5
	Inheritance from other relative	1.2
	Gift/government property	0.0
	Other	0.0
	Total	100.0
	(Number)	496

Source: Household Survey 2005.

Spatial Distribution of Households by Age

Most of the household heads were born in another house in Rashda and moved into the current residence (Table 19). They moved because of marriage, or because the previous house was old or had been demolished, but they stayed

TABLE 19 *Settlement at the current residence (household head) (2005)*

		%
From where did you move to your current residence?	From outside the village	1.9
	From within the village	98.1
	Total	100.0
	(Number)	468
Reasons for moving to the current residence	Marriage	52.3
	House demolished	40.6
	Work/study	6.2
	Other	0.8
	Total	100.0
	(Number)	497
Age of settlement in the current residence	Born in this house	10.6
	1–14	2.4
	15–24	17.1
	25–34	41.2
	35–44	20.4
	45–54	5.1
	55–64	2.6
	65 or above	0.7
	Total	100.0
	(Number)	549

Source: Household Survey 2005.

within the village. They moved when they were aged 25–44. This is clearly represented in the spatial distribution of households according to the age of their heads (Map 13). Most of the younger generation reside in the area that has been recently built.

Change of Village Lifestyle

It is clear from the information above that the development of the residential area of Rashda since the 1980s has been stimulated by movement inside the village, with the principal actors being the younger generation, motivated by marriage or a desire to improve their housing conditions.

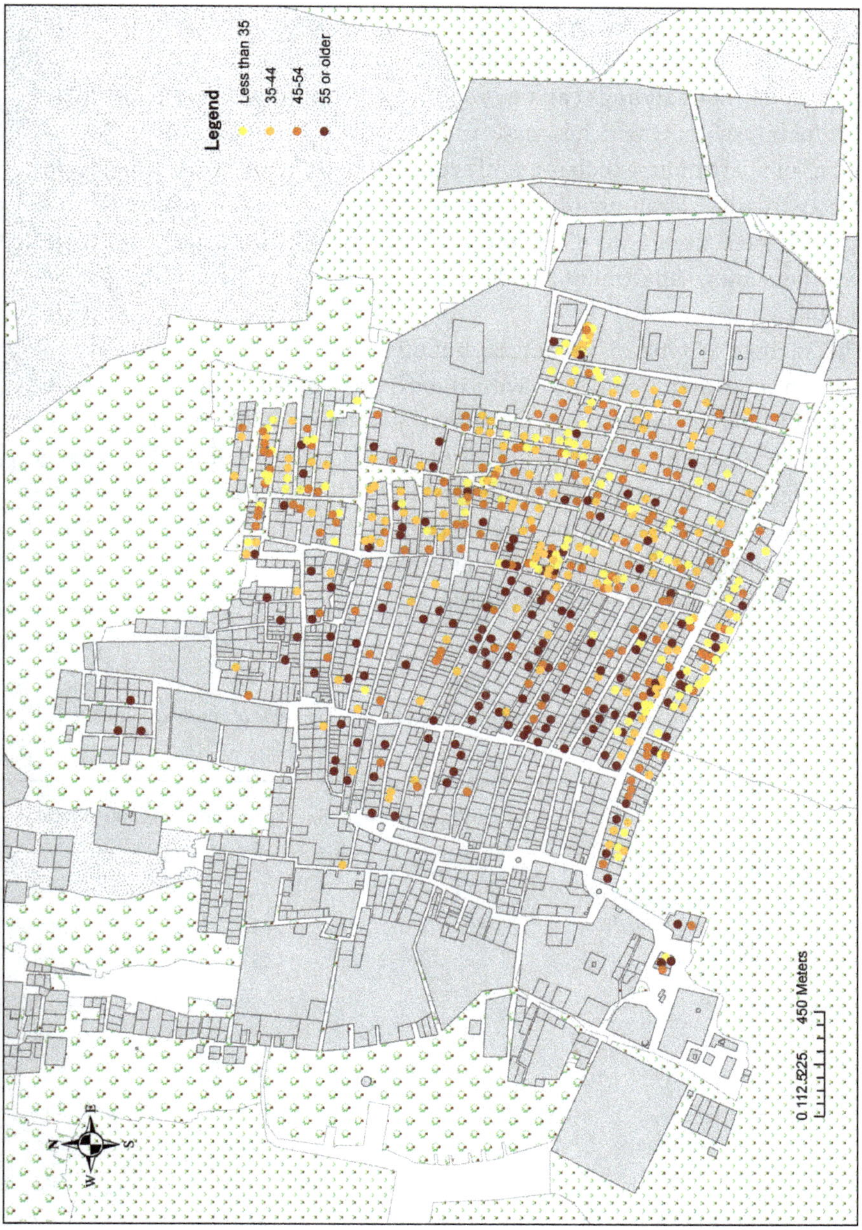

MAP 13 *Location of the households by age of the householders (unit: household head). The blocks in the southwestern residential area were not covered by the household survey. Source:* 2005 Household Survey data.

The spatial development of the residential area was accompanied by the vertical development of the residential pattern, as examined in the next chapter. Vertical development reflects a change in lifestyle among villagers. The "new" pattern of village life from the 1980s is clearly shown in the results of the field survey of the type of buildings and the commercial utilities.

The residential buildings can be roughly divided into two types. The first is the "rural house" type, which is made of mud bricks. The second is the "house" (*manzil*) or apartment, which is mainly constructed of concrete. Until 1984, the flats constructed were only of the "rural house" type. Flats constructed from 1985 to the present day, on the other hand, are mostly of the apartment type. Today, there are 50 apartment buildings with 119 occupied flats in Rashda (see Table 20).

This change in type of residential building, from the "rural house" to the "house" or apartment, coincides with the expansion of a "new" lifestyle. The apartment type of building has more than two floors, while buildings of the "rural house" type are composed mostly of ground floor dwellings. This change reflects the "new" lifestyle among villagers. Television became available in 1979, although radio broadcasting had been established in 1940. Moreover, the "new" pattern of village life is reflected in the expansion of the commercial area in Rashda, which is discussed in Chapter 8.

TABLE 20 *Number of occupied flats by year of construction and floor, type of residence (unit: flat) (2006)*

		1955–64	1965–74	1975–84	1985–94	1995–06	Total
Type of residence	Appartment		1	1	31	86	119
	House (*manzil*)		4	4	84	260	352
	Rural house	22	259	275	146	66	768
	Villa					3	3
	Total	22	264	280	261	415	1,242
Floor	Ground floor	22	261	273	146	70	772
	First floor		2	5	81	246	334
	Second floor		1	2	22	80	105
	Third floor				10	15	25
	Fourth floor				2	4	6
	Total	22	264	280	261	415	1,242

Source: Additional Sampling Survey 2006 for information on year of construction.

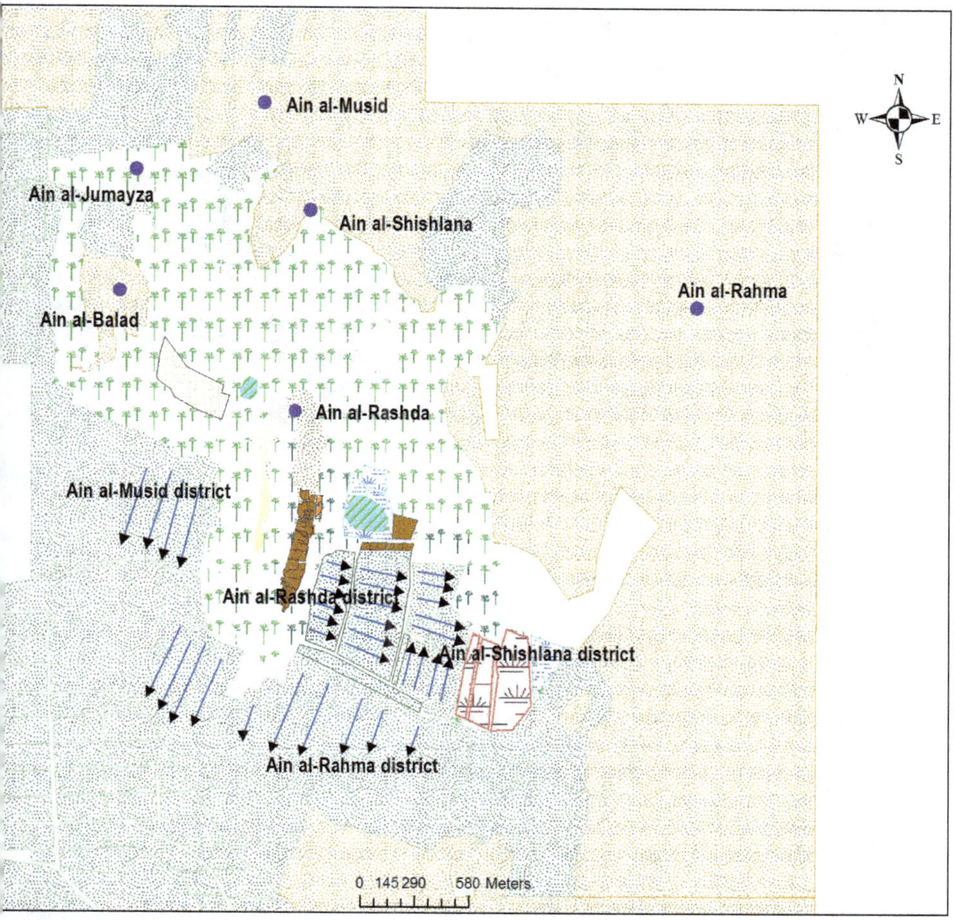

MAP 14 *Rashda at the beginning of the 20th century. Source*: Digital map based on 1/2500 scale maps (1985) and information collected by fieldwork in 2012–2013.

PHOTO 13 *Ancient office of a village chief* (maqar ʿumdiya)

PHOTO 14 *Old (right) and new (left) buildings of the Local Unit (village administration)*

PHOTO 15 *Site of the old Atiq Mosque*

PHOTO 16 *Newly renovated Atiq Mosque*

PHOTO 17 *View of the "new" Rashda with the "old" Rashda behind it*

PHOTO 18 *View of the Nazla area and the Youth Centre (bottom left)*

Excursus 2

Rashda Village Cemetery

The formation of an independent village means, in a sense, the creation of administrative borders between neighbouring villages. In this process, disputes sometimes arise concerning the reversion of sites located on the border between the neighbouring villages. The disputes become serious if the sites in dispute are symbolic of the village. The cemetery where villagers are buried is one such symbolic site for the village concerned.

The episode below concerns a dispute over the reversion of a cemetery between Rashda and Awina. Rashda and Awina are now two independent villages, but before the independence of Rashda village from Qalamun, Awina belonged to Rashda, and Rashda was a quarter (*nazla*) of Qalamun. Therefore, this episode by chance reflects the process of formation of Rashda as an administrative unit.

As mentioned in Chapter 4, Rashda was cultivated land (*ghayt*) belonging to Qalamun. All social and personal affairs, such as marriages, divorces, transaction contracts, registrations of sales and purchase, customary regulations, burials of the dead, and management of the cemetery were executed in Qalamun.

However, in 1278 (1861/62), Rashda was substantially separated from Qalamun, at which time Rashda began to be called the *nazla* (settlement), and its population register, which is discussed in Chapter 5, was produced. An interesting episode concerns this separation of Rashda from Qalamun.

Before the separation, the dead of Rashda were buried in the cemetery of Qalamun. After the separation in 1861/62, they were buried in the cemetery located in the western part of the present village of Awina. Rashda and Awina were separated from Qalamun at the same time, and Awina belonged to Rashda at that time.

This cemetery was originally the cemetery of Hindawi. Hindawi is an old and independent village, with a history that can be traced back at least 200 years, and it had a cemetery. When Rashda and Awina were separated from Qalamun, the Hindawi cemetery became the cemetery of Rashda and Awina.

Later, around 1914, during the period when Husayn Khlil was the umda of Rashda, a major conflict occurred between the umda and the village shaykhs of Rashda and the village shaykhs of Hindawi on the drinking water well (*bir al-sabil*), located in the northern part of Hindawi. The umda and the village shaykhs of Rashda wished to dig a well (*bir*) there and use its water. However, the umda of Hindawi refused to allow them to do so.

In retaliation, the umda of Rashda prohibited the villagers of Hindawi from being buried in the cemetery in Awina. Henceforth, the people of Hindawi had to find another place for a cemetery, and they founded the present cemetery of Hindawi.

The villagers of Rashda continued to bury the dead in the cemetery in Awina until 1964, when a new cemetery for Rashda alone was founded. There were two reasons for founding the new cemetery. The first was the rapid growth of population. The burial of many dead caused trouble between the villagers of Rashda and Awina. The second was the desire of villagers to be buried in a cemetery near the village.

The authority responded to this desire and specified the place for a new village cemetery according to the applicable laws and regulations. It was *Sayyid* Muhammad Hasan Sayyid Hasan who decided on the place for new cemetery. He was a barber-surgeon, and as a staff member of the Ministry of Health, he was responsible for public health and inoculation in those days. He was supported by the Committee of Social Union, which had five members from Rashda and two from Awina.

The area on the right of Photo 19 is the part allocated to the people of Awina, and the area on the left is for the people of Rashda. The white building in the distance is the Mausoleum of Sheykh Muhammad.

The Mausoleum of Shaykh Abdullah (Photo 22) is a white mausoleum along the street. This Mausoleum was on the boundary between Awina and Rashda, for the Mausoleum was a holy place that did not belong to anyone.

PHOTO 19 *Awina Cemetery*

PHOTO 20 *View of Awina from the Mausoleum of Sheykh Muhammad*

PHOTO 21 *Tombs of the umda family in Rashda*

PHOTO 22 *Mausoleum of* Shaykh *Abdullah*

CHAPTER 7

A Society Based on Water

As mentioned in Chapter 5, water was supposed to be relatively abundant in Rashda in comparison with other villages in the Dakhla oasis region. However, there was insufficient volume for cultivation, so an uncertain water supply is a condition of the social life of Rashda. Therefore, the villagers are obliged to make local rules for sharing water on one hand, and to seek government intervention in water matters on the other. This chapter describes the features of Rashda in the past as a society based on water, according to local documents. It describes state intervention in matters concerning water, the local politics of water control and the local customs, called jaala ("collaboration" in Arabic), for the digging and management of wells.

Uncertainty of Water Supply

As mentioned in the introduction, the uncertainty of the water supply is one of the three key concepts required to understand the society of Rashda. Uncertainty does not mean an absolute lack of water but instability in the supply of water that is suitable for use. In fact, the volume of usable water fluctuates by location and year.[1]

Before the 1960s, when the New Valley Project began, almost all springs and wells were shallow.[2] They could not keep pace with increasing population and areas of cultivated land, and soon dried up. As a result, attempts to identify a spring or well by name and to locate it after it has dried up are confusing because other springs or wells of the same name have since been dug in other places.

For example, Ain al-Ahmad and Ain al-Shishlana were the main sources of irrigation water for the fields of Rashda in the 19th century. Two springs of the same names are described in local documents: Ain al-Ahmad and New Ain

1 For more detail on this point, see Chapter 2.

2 According to an informant, the water source was called a *bir* (well) after permission was obtained to dig a new water source in 1910. Before that year, the water source was called an *'ain* (spring). On the subject of permission to dig a new well, see below in this chapter.

 | DOI 10.1163/9789004317390_009

al-Ahmad (Ain al-Ahmad *mustajadda*), and Ain al-Shishlana and New Ain al-Shishlana (Ain al-Shishlana *mustajadda*).[3]

A document dated 24 Dhu al-Hijja 1266 (1850) deals with the jaala (*jaʿala*) contract for the digging of the new Ain al-Shishlana, which was located to the north of Ain al-Rahma. This document regulates the distribution of rights (*amaʾil*, sing. *amila*[4]) to increased volumes of water following the digging of the New Ain al-Shishlana among the owners of the old Ain al-Shishlana and the people who dug the new well.[5] It certificates that the owners of the old and new Ain al-Shishlana were basically the same.

In the first half of the 20th century, the water discharge in Rashda decreased rapidly because of the drying up of "local wells".[6] Table 21 shows the table attached to the document dated 13 July 1937 petitioning the umda for the digging of a new well because the existing wells in Rashda had dried up.

This petition compares the quantity of water discharge between 1883, 1907, and 1937, and clearly indicates the rapid decrease in water discharge from the springs and wells in Rashda. Five springs (Ain al-Balad, Ain al-Duma, Ain Sahsah, Ain al-Rashda, and Ain Shandum),[7] which provided water to the land in Rashda had dried up by 1937. Other springs, such as Ain al-Rahma, Ain al-Shishlana, and Ain al-Mushid, which were most important in terms of water discharge volume, discharged less water in 1937 compared with previous years. As a result, the table reports that the total water discharge volume, which was 146 *qirat* in 1883, fell to 108 *qirat* in 1908 (85 *qirat* in 1907 and 23 *qirat* of water discharged from the newly drilled well in 1908) and 62 *qirat* in 1937, while the population grew from around 900 to 1,500 in 1907.

The introduction of electric well pumps to Rashda in 1941 appears to have accelerated the rate of pumping water.

---

3 On the vicissitudes of Ain al-Rahma, Ain al-Shishlana, and other old springs, refer to Chapter 10.

4 On *amila*, see Chapter 7 and 10.

5 On the jaala contract, the explanation will be found below in this chapter. The people concerned with the digging well were called the *mutajaʾilin*, which means "the people engaged with the jaala contract". They were specialists in digging and inspecting wells.

6 The "local well" (*bir ahli*) is exploited by local cultivators, who dig it in the traditional local way and own and manage it collectively. For a detailed categorization of wells, see Chapter 10.

7 On the location of Ain al-Balad, Ain al-Duma, Ain Sahsah, Ain al-Rashda, and Ain Shandum, see Map 10–2.

TABLE 21 *Table from a document dated 13 July 1937 comparing volume of usable water in 1883, 1907, and 1937*

Name of ʻain	Quantity of water in 1883		Quantity of water in 1907		Current water discharge in 1937	Report on dry springs and volumes of water from remaining springs		
	qirat	*sahm*	*qirat*	*sahm*	*qirat*		*qirat*	*sahm*
Ain al-Rahma	21	0	6	0	3	Ain al-Rahma	21	0
Ain al-Shish-lana	18	0	9	0	6			
Ain al-Musid	17	0	12	0	9			
Ain al-Jumayza	8	0	3	0	1	Ain al-Jumayza	8	0
Ain al-Balad	6	0	3	0	Dried up	Ain al-Balad	6	0
Ain al-Duma	3	0	2	0	Dried up	Ain al-Duma	3	0
Ain Sahsah	2	0	1	0	Dried up	Ain Sahsah	2	0
Ain al-Rashda	6	0	3	0	Dried up	Ain al-Rashda	6	0
Ain Shandum	1	12	1	0	Dried up	Ain Shandum	1	0
Bir al-Majunun	15	12	12	0	12			
Bir Hamam wa Mahmud	12	0	10	0	9			
Bir al-Sabil	8	0	7	0	5			
Bir al-Shaykh Abdullah	8	0	4	0	3			
Bir al-Ashara	4	0	3	0	2			
Bir Qara Birqas	8	0	6	0	7			
Bir al-Qadim	8	0	6	0	5			
Total	146	0	85	0	62	Total	47	0

TABLE 21 *Table from a document dated 13 July 1937 comparing volume of usable water in 1883, 1907, and 1937* (cont.)

Name of ʿain	Quantity of water in 1883		Quantity of water in 1907		Current water discharge in 1937	Report on dry springs and volumes of water from remaining springs		
	qirat	*sahm*	*qirat*	*sahm*	*qirat*		*qirat*	*sahm*
		Additional output in 1908				**Proportion of neighboring localities**		
		From Khandukhin	0	6		Name of village	Population	Total (*qirat*)
		From *Shaykh* Husayn	0	8		Budkhulu	around 450	107
		From Ain al-Balad	0	6		Hindawi	around 900	252
		From *Shaykh* Abdullah	0	3		Qalamun	around 1700	218
		Total of new addition		23		Maasira	around 1300	236
		Total quantity of water in 1907		85+23= 108 *qirat*		**Report on Rashda village in population census**		
						1838	900 persons	
						1907	1500 persons	

State Intervention in Water Affairs

The intervention of the state in water affairs reflects the decreasing water supply at the beginning of the 20th century. Along with the increase of population, shallow wells had dried up, necessitating that capital and technology be obtained to dig deep wells in oasis villages in Dakhla.

As a result, the state intervened in water affairs, and its influence increased every year. A memoir of the local history of Dakhla Oasis describes the process as follows.[8]

In 1299 (1881/82), wells in Dakhla were assessed by Colonel Amin Hilal Abd al-Ghaffar, but the people had previously complained to the governorate (*mudiriya*) about corruption in well assessments. Therefore, the administrator (*ma'mur*) Muhammad Islam was sent to inspect and assess the wells.

Muhammad Islam was the first administrator to visit Dakhla Oasis. The wells were reassessed, and those in Dakhla Oasis had to be legally authorized on January 1902.[9]

In 1322 (1904), a well was examined by Muhammad Sulayman of Qasr,[10] who was the manager (*wakil*) posted to Dakhla Oasis, and the well concerned was the al-Murjama well in Qasr. The examination report was drafted by the administrator (*ma'mur*) Abd al-Aziz *Effendi*,[11] who was married to Ajabana, daughter of *Shaykh* Sulayman Abu Zayd Matuq.

Sulayman organized a committee composed of village chiefs (*'umad*, sing. *'umda*). He explained that the water of the well concerned gushed too fast and was hazardous to the neighbours, who had complained about it. The committee was headed by the administrator.

The village heads were *Shaykh* Ahmad Muhammad, *Shaykh* Muhammad Husayn Abd al-Sattar, *Shaykh* Abdullah Khalafullah, and *Shaykh* Musa

8 On the water supply situation in Dakhla Oasis in those days, such as the introduction of the electric well pumps and the intervention of the state in water affairs, see Excursus 1 in Chapter 1.

9 This occurred when Rashda village became formally independent of Qalamun, and *Shaykh* Musa Ahmad Abd al-Hafiz was appointed as the umda. *Shaykh* Abu Bakr Khalil was appointed as the *shaykh al-balad* (village shaykh) of Rashda in 1902. This indicates a correlation between the change of local administration and the intervention of the state in water affairs.

10 Qasr is believed to be the oldest village in Dakhla Oasis, and was an administrative centre in the middle ages.

11 *Effendi* is the Turkish title (*affandi* in Arabic) for a non-native notable.

Ahmad.[12] They confirmed that the well concerned was harmful to the neighbours, but the complainants were not content and wanted another committee headed by the Officer (*dabit*) Sayyid *Effendi* Qutb to attend to the matter.

When the committee met, it confirmed that the first committee had answered "yes" and the second committee "no" to the question of whether the well concerned was harmful to the neighbours. Then the governor (*mudir al-mudiriya*) came to adjudicate which committee was correct, and the committee was recalled to be questioned. The members agreed to have two meetings.

They were *Shaykh* Abdullah Khalafullah (Qasr), *Shaykh* Musa Ahmad (Rashda), *Shaykh* Muhammad Husayn (Maasara),[13] and *Shaykh* Abdullah Isa. They and the administrator (*ma'mur*) were dismissed.

However, *Shaykh* Ahmad Muhammad Umar Tanbur said "we had the two meetings. The first meeting was authentic, but in the second meeting, the administrator persecuted us". The administrator denied this and insisted on remaining in his post. No investigation was conducted of this matter, and he remained in his post until the year 1329 (1911).

In summary, the authorization of wells was first performed and permission to dig wells issued by the authority (*markaz*) in 1910.[14] This means that government authorization of wells began in 1910. Previously, digging of wells had been possible without government permission.

A Society Controlled by Water

In oasis villages such as Rashda, water is the most influential factor in agricultural management. Landowners accumulate land by buying and leasing, much of which relies on obtaining water rights.

In other words, access to water resources determines the extent of the accumulation of water, land, and political power in village politics. We collected many documents on water and land purchases.[15].Here we introduce a document dated 1 Rabiʿ II 1256 (1840), which vividly demonstrates the process of land accumulation by village notables.

All praise be to Allah, alone the Lord of all creatures.

---

12 As mentioned immediately below, Muhammad Husayn, Abdullah Khalafullah and Musa Ahmad were the umda of Maasara, Qasr, and Rashda.

13 Maasara is an old village in Dakhla Oasis.

14 *Markaz* literally means "centre". It probably signifies the police.

15 They include documents on land purchases, money lending, water purchases, and so on.

[It has been agreed] by the noble, esteemed, and blessed legal court in the protected town of Qalamun located in the western Dakhla Oasis, in the presence of its custodian our master and Lord the legal administrator,[16] who will affix his noble signature and seal above, may Allah grant him his eternal blessings and make him victorious by Prophet Muhammad, the elite among his creation ... [that] the honourable *Shaykh* Tantawi and his brothers: Muhammad, Abd al-Tawab, and Husayn, sons of Abd al-Hafiz Husayn, and the sons of their late brother Nasr, i.e., Humam and his brother Shihab, all originating from Qalamun in Oasis, are co-renting exclusively with their own money—on the understanding that [the rental fee] is to be divided into five portions, one portion to be borne by Shihab and his brother while the remaining portions [four portions] will be borne by *Shaykh* Tantawi and his brothers ... [from] (that is, co-renting from) *Shaykh* Hasan and his brother Abd al-Rabb, sons of the late Abd alSaud, who come from Qalamun, each acting in his individual capacity, while Hasan is acting in his individual capacity as well as standing proxy by power of attorney for his sister, a respectable lady of full age and capacity, named Ghawat al-Thabit, as witnessed in a legally certified manner by *Shaykh* Shaarawi, son of the late Masud Qaymi, and *Shaykh* Muhammad, son of the late Ibrahim Abd al-Nazir, [knowing that] Mrs. Ghawat is the legal custodian of her children, her son Makki and his sibling Turkia who were born from her marriage with her late husband Ali Makki ... who are yet to come of age and are by necessity entirely dependent on their mother for their sustenance and livelihood, as duly acknowledged by our lord the legal administrator. The above-mentioned lessors [*Shaykh* Hasan and his brother Abd al-Rabb] are leasing [to the above-mentioned leaseholders] a property that is duly owned by them and by those for whom they stand proxy as a rightful legacy; this property is the entire allotment (*hissa*) named Ghytaniya, consisting of a plot of land planted with various types of palm trees found at the Ain al-Rashda orchard, which is located in the Northern district of Qalamun city. The four boundaries of this plot of land are as follows: the south is (the land of) Masud Hasan; the north is (the land of) the above-mentioned leaseholders; the east and west are sand hillocks. [The lease also involves] the lessors' entire share of drinking water ... [quantities of this share of water are specified according to local units of measurement,] which is provided by a nearby well and water spring in the above-mentioned Ain al-Rashda. The lease period has been fixed for thirty *ʿaqd* (contract), with each *ʿaqd* equalling three

16 This official is the equivalent of a *qadi* (judge).

> years, which means that the overall lease period totals ninety years starting from the date on which this [contract] has been drafted, noting that each year in this period is to be taken individually as a single year; that is, it shall not overlap either with the following year or the previous year ... the years of the lease must be legal (i.e., counted in accordance with the Islamic calendar) and should be consecutive; the annual rental fee in Egyptian *nisf fidda* totals eight thousand five hundred and twenty *nisf fidda.*[17] The transaction must be settled on the spot upon the signature of this contract; that is, payment of the total sum of eight thousand five hundred and twenty *nisf fidda* entailed as follows: three thousand five hundred and forty *nisf fidda* to be paid as a rental fee for the water and five thousand seven hundred *nisf fidda* as a rental fee for the date palms. The total amount of the rental fee has been duly handed over [by the leaseholders] to the above-mentioned lessors upon the conclusion of the present contract between the two parties in the presence of witnesses who have duly observed the process of this transaction in compliance with the precepts of Islamic law (*shari'a*). As such, the leaseholders have fully acquitted themselves of the above-mentioned rental fee by handing over to the lessors the entire amount of the rent in cash. Thereupon, the lessors have legally (in compliance with the Islamic law) and voluntarily given up possession of their property in favour of the leaseholders, who have accepted their shares as previously detailed in compliance with the precepts of the Islamic law declaring these shares as their own and claiming their ownership rights to them. This transaction has been duly acknowledged and testified by qualified witnesses and Allah is the best of witnesses. It has been agreed that should any legitimate issue calling into question the validity of the lease arise, it is up to the lessors to refund the equivalent of the rental fee to the leaseholders without indemnity, compensation, or counterclaim. It has also been acknowledged by the witnesses to the power of attorney deed that the proxy [*Shaykh* Hasan] has duly apportioned the dues of the principal [the person he is representing; i.e., his sister Ghawat al-Thabit], thus discharging himself from all liabilities. In witness thereof, the undersigned have affixed their signatures on the first of Rabi II, 1256 (one thousand two hundred and fifty six).

This document, which is categorized as a lease contract, deals with a long-term lease of land by the family of the future umda of Rashda, who had approximately 80% of the land in the fields of Rashda. The land transaction is called

---

17 The *nisf fidda* literally means "half silver", and was a unit of currency.

an *ijar* (lease) in the document. However, it is substantially a contract for purchase, because the lease period is 90 years.

A Society Based on Water Sharing

The owners of water undertook to cultivate their land, putting the farmers under their influence. However, this undertaking could not be successful under conditions of uncertain water supply and limited capital and technology. They had to collaborate with other owners of water and farmers working on their land; there was too little water for unlimited use. In fact, as mentioned above, the shallow spring or well would dry up shortly thereafter.

The jaala (*jaʿala* in literary Arabic) custom is a means of collaborating on water affairs that compensates for the shortage of capital, labour, and technology. The formation of the village of Rashda was itself deeply connected with the sharing of water resources. As discussed in detail in Chapter 5, the core of migrants from Qalamun to Rashda was composed of those who shared the water from Ain al-Rahma. They were called *shurakaʾ* (partners). Moreover, the irrigation system has basically been managed by units of irrigation districts based on the sharing of common springs or wells, as discussed in Chapter 11.

A jaala is a legal act that can be used to contract labour to produce a result if the amount of labour required cannot be determined (Peters 2008). According to Peters, the jaala contract is an exception to the strict rule that in synallagmatic contracts, the provisions must be known and well defined. This means that the jaala is a contract based on customary law (*ʿurf*).

We collected many jaala documents in the possession of village notables during field surveys.[18] All of them are concerned with the labour for digging and managing wells. Therefore, the jaala documents are precious source materials for understanding how societies based on underground water were managed in an oasis village.

Jaala for Digging a Well

A document dated Dhu al-Qaʿda, 1326 (1908) provides detailed information on the procedure for the digging and completion of a well by jaala contract. The

18 Almost all of the jaala documents that we collected during the field surveys are related to Qalamun, because Rashda is a modern village founded in the 19th century. As mentioned in Chapter 5, Qalamun is the village to which Rashda belonged before its independence as an administrative unit. The oldest jaala document that we collected is dated 23 Jumada II 1201 (1787).

document consists of two parts. The first is a table specifying tasks to be accomplished and the shares of water rights as a reward for these tasks in the digging the well, and the second is the summary of the contents of the jaala document. The following is a translation.

> Report on the distribution of (the water rights) of Ain al-Shawadla, located in the north of the village of Qalamun, which belonged to its *zimam;* its contents are explained below. It was drafted on 20 Rabiʿ I, 1331 (1913).

TABLE 22 *Distribution of the water rights of Ain al-Shawadla*

	Amila	*Qadam*	*Qirat*
Source of well (*asl al-bir*)	75	0	
for the rent of a boring derrick (*dulab*)	7	15	
(for the cost of) sant (acacia wood) for building wood pipe (*zamamir*)	6	0	
(for the wage of) 13 laborers for digging	6	0	
for the work of *najjarin*	2	15	
(for) the inspection	2	0	
(for)the digging and amending tools, and oil and engine	2	0	
(for) the cylinder for working the boring derrick (*dulab*)	1	0	
(for) the waqf (*waqfiya*) and the secretary work of Abd al-Hamn Hasanayn	1	0	
for the transportation of the boring derrick (*dulab*) and the official permission	2	0	
the distribution (of water rights) among the *mullak* (owners) of the spring is done according to the cost (*taklif*) mentioned	45	0	
The total is 75 *amila* according to the rule above.	75		

Note: The *sant* tree (*Acacia nilotica*) grows abundantly around the oasis, and its wood is extraordinarily durable in water.

> On the above date, the undersigned sharers of the Ain al-Shawadla well have agreed on a consensual basis on the tenor of the present report, recognizing that each sharer is committed to provide, in his own capacity and in accordance with his ownership ratio, the tasks and equipment requested from him. Should any sharer fail to secure in a timely manner all the required tasks and equipment, he must renounce his share and cede it to the persons who would carry out the work for him. As such, if a

> person or a group of persons carry out the tasks of a defaulter, the latter will have to forsake his entitlement in his favour and he will not be allowed to seek legal recourse to claim back his portion, based on mutual agreement without any compulsion of any kind on either side. All the undersigned sharers have entered into this agreement freely and voluntarily on a consensual basis as from the above-mentioned date.

This document was drafted in 1908. As mentioned above, the supply of usable water in Rashda was decreasing at the beginning of the 20th century because of the drying up of the "local wells". Therefore, this jaala contract was concluded for digging a new well.

This document is not complete, because the names of people who were involved in the jaala contract are not mentioned in this document. It is supposed that there was a document or memorandum in which the names of people concerned with the jaala and their shares of water rights were mentioned.

The investors who finance the digging of a new well are called *mullak* (owners). This is why they were the joint owners of the well that already existed and was probably located in the area irrigated by the water of Ain al-Shawadla. They were apparently planning a joint jaala project to dig a new well, because the water from the old well was decreasing or had dried up. Their names are listed in the additional document, because a sentence is found in the document to the effect that "the distribution (of water rights) among the *mullak* (owners) of the spring is done according to the cost (*taklif*) mentioned (in another document)".

In this document, the specialists in charge of digging or managing a well, such as diggers, carpenters, scribes and inspectors, as well as their shares of the water rights that they received in return for their tasks and work, are listed. The criteria by which water rights were shared among the people were stipulated in the jaala. However, it is supposed that the water rights were distributed according to the rate of investment for the owners and the kind of work and services supplied by the specialists.

In summary, this is a document that certifies that the digging of a new well was planned in the framework of a jaala and that all the people concerned agreed to share the water rights with the specialists in return for their contribution.

Jaala for Managing a Well

However, the jaala contracts cover not only the digging of new wells but also the management of existing wells. The jaala contracts for the management of wells are basically drafted in the same style as those for digging. The management of wells includes various kinds of tasks and work, of which the main ones are repairing, cleaning, and inspecting canals. We collected a document dated Dhu al-Qaʿda, 1326 (1908), which deals with a jaala contract for the repair of a canal called Ain al-Najjarin in Balat. Its content is as follows.

> Jaala contract
>
> All the owners of Ain al-Najjarin are present, and they are the honourable Hamuda Mustafa al-Najjar, the honourable Sayyid Sultan, his brother *Hajj* Muhammad, the honourable *Hajj* Mansur Saad, the honourable Hamuda Farraj, his brother Abdullah and the honourable Zanqur Nasr Muhammad. They agreed (*tawafaqu*), consented (*taradu*), and entrusted (*jaʿalu*) one of the partners (*shuraka*ʾ), Zanqur Nasr Muhammad, with the work related to the Ain concerned [Ain al-Najjarin] as follows. The canal, which they dug, passes from the Ain from west to south until the old place of *saqiya* (*maghtash al-saqiya*), and the water flows between two banks to the place mentioned.[19] A part of the canal suffers from sand. Therefore, he [Zanqur Nasr Muhammad] defends the Ain by fencing the canal with grasses (*halaf*).[20] When he completes his work as regulated and experienced persons (*arbab al-khabra*) confirm it, he shall have 22 *amila* in return for his work. He has the right to 11 *amila* of it to dispose of as he wants. The remaining 11 *amila* are given to him as long as he works. These 22 *amila* remain to be delivered by the rotation (*dawra*) [of the water] of the Ain, which is 44 *amila*. However, when it is certified that the work is not done as regulated or the work cannot be done, he has no right and is not entitled to claim the right according to the contract. They agreed, consented, and confirmed this personally, and the witnesses also confirmed it. Then, the sant tress on the Ain and the canal are now divided among landowners,[21] excluding the cover of the roof (*habbaka*) [of the Ain]. This [contract] was drafted exactly according to the necessary procedures and was issued on Dhu al-Qaʿda in.1326.

19 *Saqiya* is a device for lifting water for irrigation, which uses buckets, jars, or scoops fastened to a vertical wheel.

20 *Halaf* are the long, strong grasses (*hashaʾish*) used for fencing from sand.

21 *Sant* is a very strong tree that grows around the water springs, and its wood is used for furniture, tools, and other purposes.

The people concerned with the jaala contract are called the *mutaja'ilin*, but at the same time, they were called the *musharikin* (sing. *musharik*) or shuraka' (sing. *sharik*), which means that they were partners. This shows that their relationship was considered to be a kind of partnership (*musharaka, sharika*).

In our view, it is essentially instructive that the relationship of those who were concerned with the digging and management of a well were called members of partnerships, because this implies that the jaala related to the well formed a kind of water association or union.

PART 3

Rashda Today

CHAPTER 8

Rashda as a Community

Rashda was registered as an independent village at the end of the 19th century, as was mentioned in Chapter 6. Rashda today provides many services to villagers as a community. It is, first of all, an administrative unit, and has official functions related to its administration. This chapter aims to provide basic information about the official institutions concerned with the public functions of the administrative unit: location and the administrative system, the arrangement of village space, buildings for the public, commercial and residential facilities, and public services provided by public and non-governmental organizations.

Location and Administration

Location

Rashda today is a village in the District (*markaz*) of Dakhla in the New Valley governorate. Rashda is the administrative centre, and it is usually called the "mother village" of the Local Unit for Villages of Rashda (*al-wahda al-mahalliya li-qura al-rashda*). The present Local Unit of Rashda is composed of Rashda, the mother village where the administrative office is located, and the two satellite villages of Ain Abu Othman and Duhus.

Rashda is located 10 km north-west of Mut city, the administrative centre of Dakhla District. The Local Unit of Rashda adjoins the mountainous region to the north and shares a border with Mut city in the south, with the Local Unit of Hindawi in the east, and with the Local Unit of Qalamun and the Local Unit of Jadida in the west (Map 15).

The Local Unit of Rashda is the largest unit in Dakhla District after Mut in terms of population. According to an estimate by the Local Unit, the total number of households was estimated to be 1,789 in 2014, with a total population of 6,862 in 2014. The population of Rashda itself was 6,292 in 2014.

Every morning, minibuses take people to Mut, the central city of Dakhla District. Until 1980, there were only three cars (*jeep qadim*) that provided the shuttle service between Rashda and Mut. Now, five microbuses of five or six seats offer a shuttle service between Rashda and Mut from 7 am until 6 pm every day. There are also taxis from Rashda to Mut, and even a microbus that has run between Rashda and Cairo since 1995. Now the villagers can either take

 | DOI 10.1163/9789004317390_010

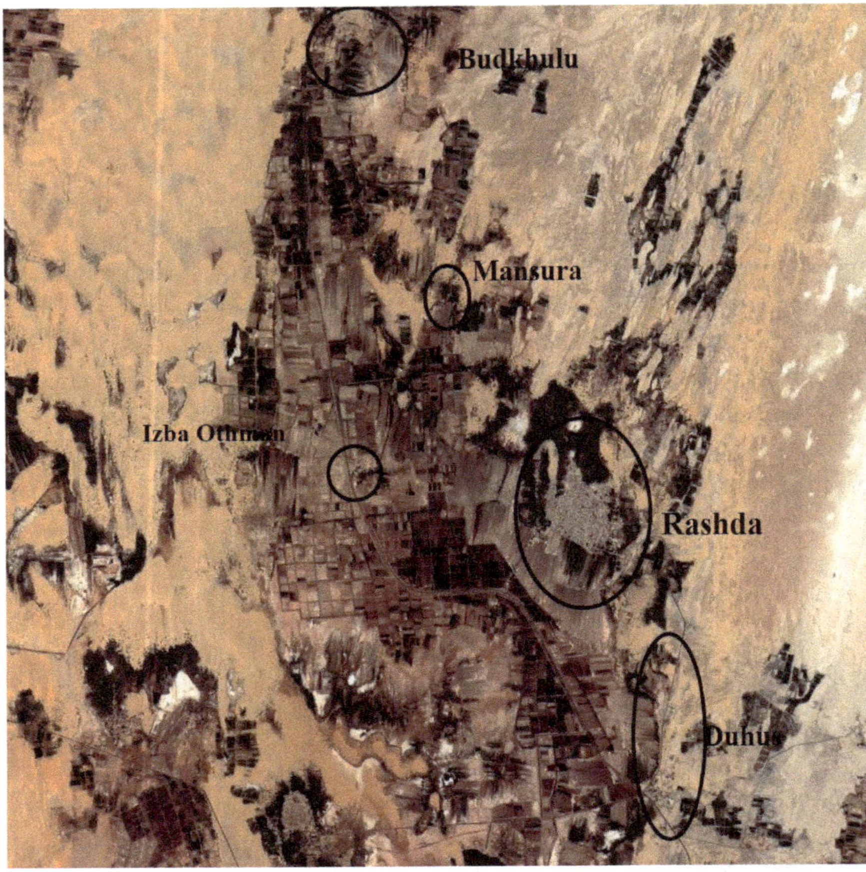

MAP 15 *Rashda and its associated small settlements*

a microbus direct from Rashda to Cairo or take a "Superjet" bus at night from Mut to Cairo, a service that began in 1993.

Administration

The Local Unit of Rashda manages village administration. The present administrative framework of Rashda was established in the 1960s under the socialist regime of Gamal Abdel Nasser. In Rashda, like other villages in the governorate, a committee of a socialist union was established in 1963. It was composed of about 20 villagers: 11 members from Rashda and nine members (six teachers and three elderly people) from Awina village, which was integrated into Rashda village.

In 1970, the Local Unit of Rashda was established, and it is one of the oldest local units in the Dakhla District. At the time of establishment, it was composed of four villages: Rashda village (the mother village), Hindawi village, Awina village, and Budkhulu village. In 2002, the local units were rearranged, and Hindawi became an independent local unit with Jadida village. Budkhulu village formed a local unit with the two small settlements of ʿArab and Mansura.

The Local Unit of Rashda, like other villages in Egypt, has an executive council, led by a unit chief (*raʾis al-wahda* or *raʾis al-qarya*), composed of the representatives of the major ministries: Interior, Education, Social Affairs, Health, Agriculture, and Housing, and the Agricultural Bank. The unit chief is appointed by the governor and is responsible for supervising the executive and administrative agencies working for, and providing services to, the villagers within the local unit.

One such service concerns sewage, which is commonly an inconvenience in Egyptian villages. In 2004, the local unit planned a sewerage project to service the entire village of Rashda. Garbage is collected systematically every day, using a truck from the local unit.

The local unit also has a People's Council (*majlis al-shaʿb*) elected by the villagers. This council is composed of 24 members, of which 12 are elected by the villagers and 12 are representatives of the major ministries mentioned above and appointed by the unit chief. Its main role is to supervise the implementation of activities under the executive council and to propose village development projects based on the villagers' requirements. The council is composed of several subcommittees such as the agricultural committee and the housing committee. Officially, when a villager has a problem that needs to be resolved concerning social and land conflicts or irrigation matters, he contacts the people's council.

However, the people's council has been under suspension since the 2011 revolution. In such cases, when the council does not function, the unit chief nominates one or more villagers who are expert in the domain, such as elders (*kubar*). For instance, in 2014, one villager asked the unit chief to resolve a conflict with his neighbour regarding the border of a house. This should have been resolved by the subcommittee of housing, but because it was not operating, the unit chief asked an elder to intervene and resolve the problem.

The traditional system of village administration, headed by an umda and supported by a Customary Council (*majlis ʿurfi*) officially ended with the establishment of local units in 1970. Nowadays, the umda (village chief), who headed the "traditional" administration system, has been appointed, rather than elected by villagers, since 1994. Officially, an umda is responsible for security. In other words, it may be said that his role is limited to security issues.

By law, there should be an umda when there is no police station in a village. There is no police station in Rashda; instead, the umda and watchmen (*gafir*) are responsible for security issues. Watchmen are a group of seven to nine men, and like the umda, they are public servants. There is also one *shaykh al-balad* (village shaykh) who supports the umda in maintaining security.

As to the Customary Council, it played an important role in village management until 25 years ago, but it has been replaced by the popular council, according to the staff of the local unit.

Nowadays, the Customary Council does not exist in an institutionalized way. The villagers may consult the elders (*kubar*, or sometimes called *shaykh*) when they need help in solving problems within the village. The elders are four people from four old families: Ababida, Jamal Din, Khair Allah, and Abbud. The shaykhs of the two old mosques, Abu Bakr mosque (which is perceived as being attached to the family of Abd al-Hafiz) and Nur mosque, are also important people for consultation.

However, it is noticeable that the Customary Council and People's Council are not opposed to each other. Rather than an unofficial or traditional system that has been changed to an official or modern system, the latter developed by the inclusion of the former. This point can be understood in the composition of the popular council.

Table 23 shows the functions of each member of the People's Council. We collected this information in a government well irrigation district in Rashda. There we asked the farmers whether they had any relatives or acquaintances on the People's Council, and if so, what their roles were. The answers to this question reveal two interesting features of the current People's Council.

One is that the members of the People's Council include those who traditionally composed the Customary Council, and they hold important functions such as the management of wells and distribution of water and land. Thus, in a sense, the People's Council is a renewed version of the Customary Council.

Another interesting feature is that the family still seems to play an important role in village politics, because the members of the People's Council come from old families with elders on the Customary Council. Except for three persons, the members belong to old families.

Thus, in practice, the border between official (*rasmi*) and unofficial (*ʿurfi*) is not clear. In reality, many formal and informal bodies are involved in decision-making in village life. Table 24 illustrates how different bodies are concerned with different issues in village life. In land disputes of primary importance such as one between villagers, such as the land owned through "putting his hand" (*wadʿ al-yad*), which means literally the possession, an elder intervenes, and if the dispute cannot be resolved by the elder, the umda or the police intervene. A land dispute of secondary importance, such as one between brothers or relatives, is essentially resolved by a family chief (*shaykh al-ʿaʾila*) or the

TABLE 23 *Roles of the members of the People's Council* (majlis al-shaʿb) *in Rashda* (2010)

Family (aila)	Role in the village
R	*Shaykh* (chief) of mosques, chief of committee related to religious issues, resolving social issues
Ab	Inheritance, resolving conflicts, history of the village
F	Water schedule registers of local wells
F	Member as the head of Zakat committee
Ab	Member as representative of the agricultural cooperative, resolving conflict
J	Supervisor of governement wells as representative of the irrigation department
JD	Resolving conflicts and problems, land sale contract
AH	*Kubar* (elders), representative of the charitable association and the Zakat committee
AH	*Shaykh* (chief) of mosque, advises in resolving problems
AH	Umda, responsible of customary council for resolving problems
AA	Director of the Social Unit (Ministry of Social Affairs)
Abd	*Ma'dhun* (contract of marriage and divorce), resolving the marital problems
Abd	*Shaykh* (chief) of mosque
U	Government land property, division of land *zimam*
U	Promoting religion, Azhari school, resolving problems
GB	Reolving conflicts
AH	Representative of education, resolving problems related to education
Mb	Land sale contract, responsible for water schedule in Government Well No.1 Irrigation District
Mu	Resolving conflicts, financial support, donation

Source: Information collected by informant in 2010.

chief of the village quarter. Inheritance is solved essentially by elders, and family problems such as divorces are resolved within a family by relatives and the family chief. Water disputes are resolved by the chief of water, or by the umda or police.

The result of the household survey in 2005 also reveals the coexistence of different solution mechanisms. In Rashda, 550 household heads were asked "who do you prefer to ask to resolve disputes" concerning "land related matters", "inheritance", and "family problems". For land-related matters, 39.8% of household heads replied "the umda" followed by "the family chief" (35.3%), and "villagers" (18.0%). For issues related to inheritance, 72.2% of household heads replied "the family chief" followed by the umda (15.8%). For family

TABLE 24 *People/institutions relied upon to resolve problems (2010)*

		Umda	*Shaykh al-balad*	Elders (*kubar*)	Police	Supervisor of irrigation in government well, water chief in local wells	Family chief (*shaykh al-aila*)	Relatives
Land	Primary importance	•		•	•			
	Secondary importance		•				•	
	Inheritance			•				
	Family problem						•	•
Water	Primary importance				•	•		
	Secondary importance	•				•		

Source: Information collected by informant in 2010.

problems, 84.4% of household heads nominated the family chief followed by the umda (8.0%).

Residential Space in Rashda

Rashda is a modern village that does not have the traditional landscape of an old village in the Western Desert. The village is not encircled by walls, and its residential districts are not divided into quarters (*darb*). Today, Rashda is composed of five residential areas as follows (refer to Map 12 in Chapter 6).

1. The "old" Rashda, located on a sandy hill, which nowadays is composed mainly of demolished houses. Some houses are still occupied, but most are abandoned.
2. The "new" Rashda, which extends east of the old area, between three principal streets (*shariʿ*) of the village: Midan Street, Tijari Street (Commercial Street), and Masjid Rahman Street (Rahman Mosque Street).
3. Izba Shamrukh, which is located at the eastern edge of the village, between Masjid Rahman Street and Amel Street, which is at the end of the village. Izba Shamrukh comes from the name of the family that first moved into this area.
4. Nazla, which is located in the northern part beyond Nazla Street.
5. The area called Qara, which is located behind the old area, where some 20 households reside.

Buildings for Utility

In the areas numbered 1 to 4 above, 1,102 buildings were identified as occupied in the 2006 sampling survey. Among them, 132 buildings are for utilities. "Utilities" refers to public and commercial facilities usually occupying a building or part of a building. As mentioned in Chapter 6, Rashda was formally recognized as a village and became an administrative unit in the early 20th century. With this legal establishment, Rashda began to acquire some basic facilities. For instance, Atiq Mosque was established in 1917, and Rashda Primary School in 1922. However, their buildings are no longer used: the former was renovated (completed in 2013), and the latter has been transformed into the Food Rationing (*tamwin*) Office.[1] The third oldest building is the Local Unit

1 A Food Rationing Office exists in Rashda, and offers qualified individuals supplies of subsidized foods and goods such as oil. The Food Rationing Office in Rashda was established in

office founded in 1970, but its original building is currently used as a kindergarten. It was moved to its current place in 1981 and renovated in 1991.

A second wave of construction of modern utilities came in the 1960s during the period of the New Valley project. Among the buildings currently in use, only four were constructed before 1970. According to the establishment survey, all are in public use. The oldest building is the umda (village chief) office (*maqar al-ʿumdiya*), established in 1934. The other three were all established during the 1960s: the Primary School and Agricultural Co-operative in 1962, and the Health Unit in 1964. It may be added that some of the basic infrastructure was constructed during this period: the asphalted road to Mut in 1967, and electricity (six hours per day from 6 pm to midnight) in 1969.[2]

The other basic utilities and infrastructure were provided in the 1970s. Three buildings were constructed: Abu Bakr Mosque in 1975, the Youth Centre in 1976, and the Post Office and "Central" Rashda (Telecommunications) in 1979. In the latter half of the 1970s, the water network became available; the well for drinking water was constructed in 1975, followed by the public water network in 1978.

In the 1980s, more buildings for social and cultural utilities were constructed: the Rashda Social Development Association (NGO), which runs kindergartens, in 1980, the gathering place of Abu Bakr Mosque in 1981,[3] Nazla Mosque and Rashda Preparatory/Secondary School in 1982, the Cultural Centre in 1985,[4] the Veterinary Unit and Agricultural Bank (Village Bank) in 1987,[5] and Mubarak Primary School in 1988. The Social Affairs Unit, which provides monetary services to the needy (mainly to widows and orphans), was established in 1990, and another primary school (Amel Preparatory School) was established in 1995.

1997. There are five shops in Rashda selling subsidized food. There is one bakery in Rashda selling subsidized bread. Each person has the right to buy five loaves of bread per day.

2 Radio broadcasting became available in 1940, and television in 1979, as mentioned in Chapter 6.

3 "*Dar al-munasabat*" refers to gathering places used for occasions such as weddings and funerals.

4 The Culture Centre in Rashda functions as a library and offers lectures, group discussions, and contests.

5 The Village Bank (or Agricultural Bank) is defined as the financial unit derived from the Principle Bank for Development and Agricultural Credit (PBDAC) to provide financial services for farmers who are members of the Agricultural Co-operative. The Village Bank in Rashda was established in 1987.

Buildings for Commerce and Residence

Today, the buildings for public utilities are mainly located at the entrance of the village (refer to Map 12 on page 110 and Map 16). The bus stop is located on the square (*midan*) there. Five mosques and schools are located on the edges of the village. The cemetery is at the south-east edge.[6]

Shops

Shops are found mostly along the main street of the village, called "Commercial Street" (Tijari Street). Land plots used for storage and small workshops (*warsha*) are located in the south-eastern area behind Izba Shamrukh and in the old area.

No commercial utilities were built before 1980. The oldest grocery shop (*baqqala*) that still remains in Rashda dates from 1984. In the following year, 1985, another grocery shop and a barber-shop were constructed. The number of commercial utilities has increased since the 1990s. In 2008, there were 112 commercial shops, including 36 grocery shops, three butcher's shops, four grain mills, four clothes shops, and nine barbers. There is also some manufacturing, mostly for housing, such as five factories making windows and doors (Map 16).

Houses

There are a total of 970 residential buildings according to the sampling survey. They are all located in the "new" Rashda, Nazla, and Izba Shamrukh areas. As mentioned in Chapter 6, there are only 15 occupied residential buildings that were constructed before 1964. Construction of residences began after 1965 and has continued constantly until the present day (Map 17). They are located at the eastern side and the northern part of the village, and most recently built residences are located in the Izba Shamrukh area, which still contains vacant land and orchards.

The first house constructed in the "new" area in 1961 is that nearest the old area, and the second is in front of Umar Ibn Khattab Mosque along Commercial Street (Tijari Street), built in 1962. In the Izba Shamrukh area, the oldest house is one constructed in 1975. Its owner's family recalls that at that time, their house was the only house in that area, and until the beginning of the 1980s, it was surrounded by agricultural land.

---

6 The old cemetery is located in Awina village, located at the eastern end of Rashda. The new cemetery was constructed around 1920, after the separation of Rashda from Awina and Hindawi. See Excursus 2 in Chapter 1, and Chapter 6.

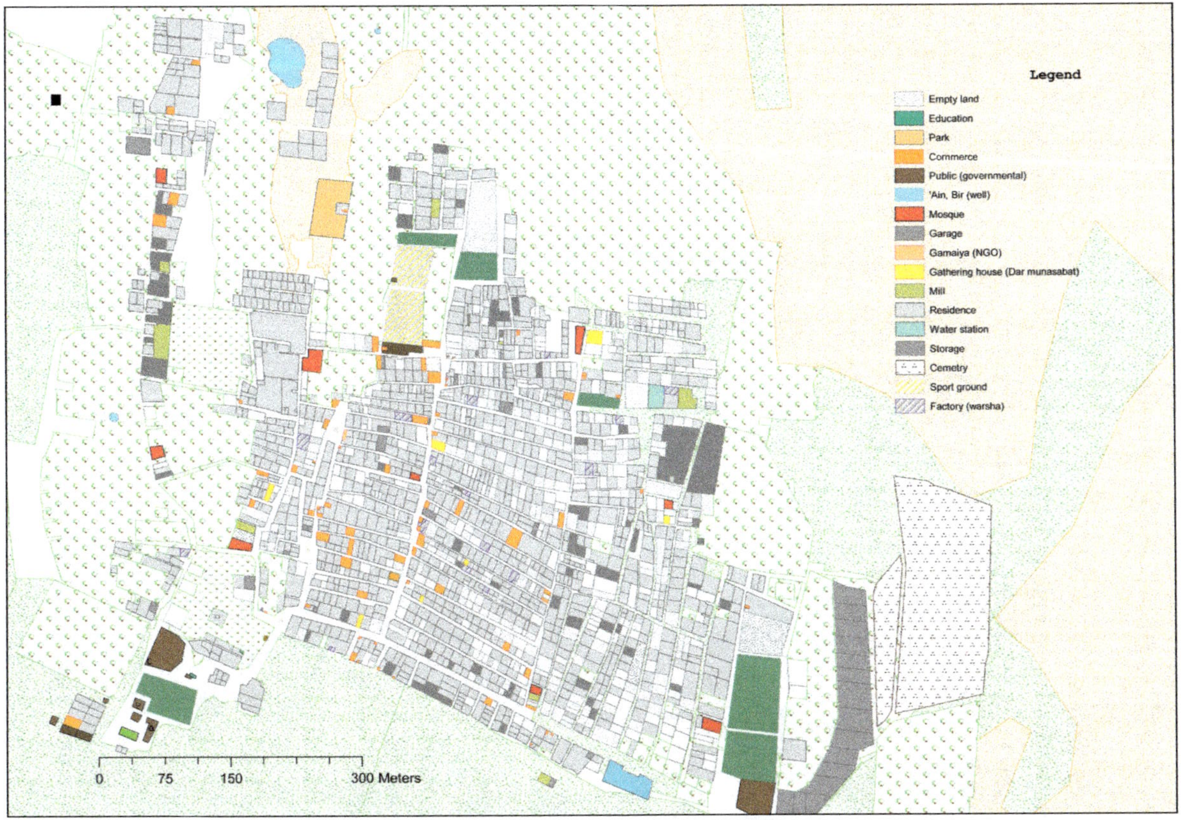

MAP 16 *Location of the buildings for public and commercial utilities (unit: building) (2008). Source*: Kato and Iwasaki 2008: 35.

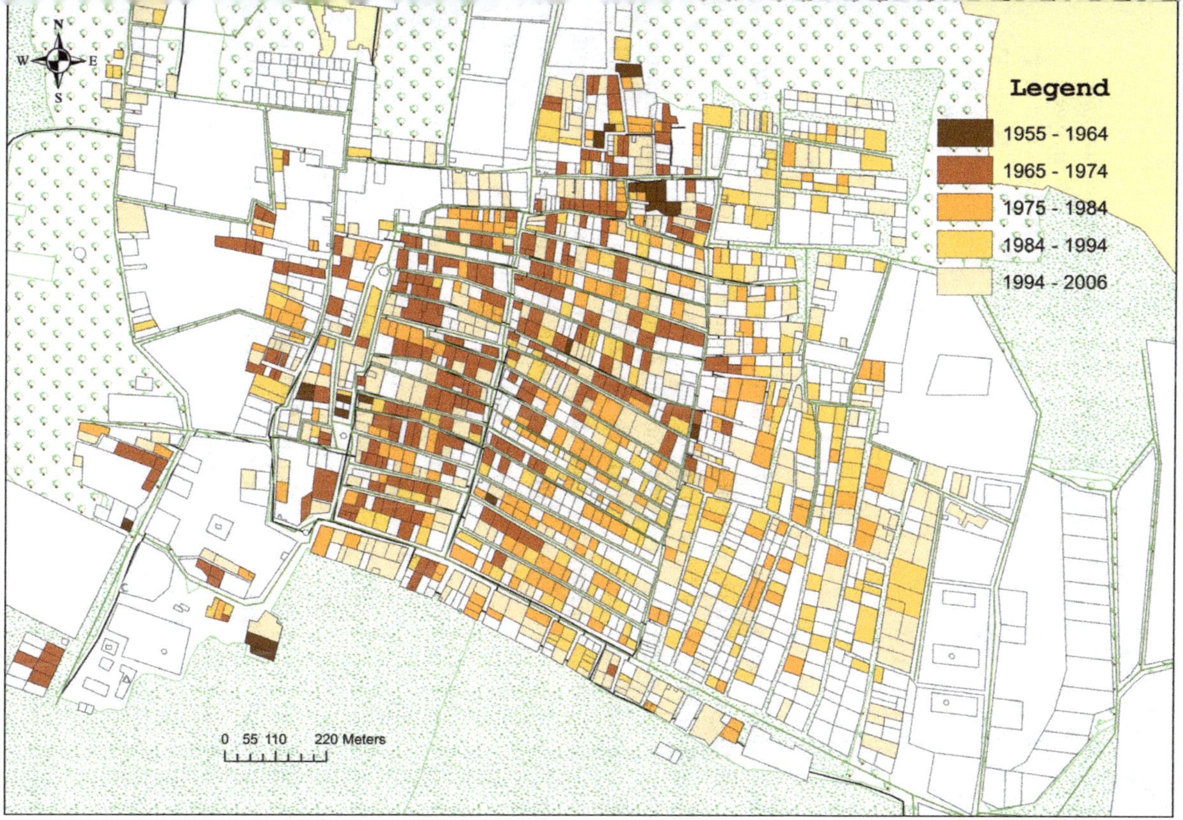

MAP 17 *Occupied residential buildings by year of construction (unit: building) (2008).* Notes: *1. Year of construction refers to that of the ground floor for buildings composed of more than one floor; 2. The year of construction displayed on the map refers only to occupied buildings used for residential purposes. The blank spaces are non-residential buildings, those under construction, or vacant land. Source*: Kato and Iwasaki 2008: 28.

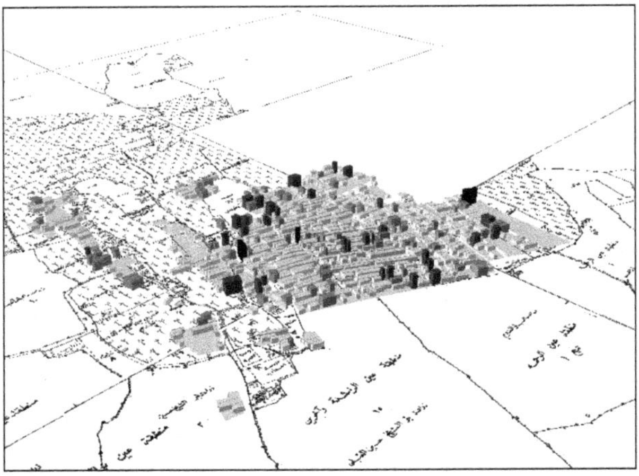

MAP 18
Building heights by floor (unit: building) (2008). Note: the height of each building is estimated using information on the number of floors in each building.
Source: Kato and Iwasaki 2008: 30.

The horizontal expansion of the residential area mentioned above was concurrent with the vertical development of the residences. This is apparent from the construction of flats by floor. In total, 1,242 flats are occupied by households. Most of these occupied flats are located on the ground floor, because 69% of the residentially occupied buildings have only a ground floor. The tallest occupied residential buildings in Rashda have four floors; there are four of these buildings. However, it should be noted that 27% of flats are located on the first floor, 8% on the second floor, 2% on the third floor, and 0.5% on the fourth floor.

In terms of years, most of the construction took place at the ground-floor level until 1984. From 1985, and notably from 1995 until the present, the construction of first, second, and third floors accelerated. It may be noticed that the construction of upper floors was especially common in the "new" area between Rahman Mosque Street (Masjid Rahman Street) and Commercial Street (Tijari Street). In the Izba Shamrukh area, which is the most recently developed area, the buildings have a single storey (Map 18). Thus, it can be pointed out that the development of residential space has taken place not only horizontally but also vertically in the past few decades.

Public Services in Rashda

Organizations for Public Services

In Rashda, besides the Local Unit, many organizations provide public services of various kinds. The following play important roles in this field: the Social

Unit, the Agricultural Co-operative, the Youth Centre, schools, the Health Unit, and religious facilities.

Social Unit

The Social Unit is an administration office of the Ministry of Social Affairs, established in each village. The Social Unit in Rashda administers such items as retirement pensions, soldiers' pensions, financial assistance, and the Productive Family Project. It had 12 staff members and workers in 2004.

The Soldiers' Pension (or Servicemen's Families Pension) provides aid for soldiers and the families of those who served—whether a husband, a father, or a son—to develop and increase their income. Moreover, cash aid is provided to families of military personnel who have been killed or civilians who have been injured as a result of military operations or land mine explosions.

The Productive Family Project is a social project that started in 1964 in Egypt. It is intended to develop the economic resources of people in low-income categories and young graduates, providing cash aid as well as vocational training and marketing services. In addition, the Social Unit provides financial assistance (monthly cash aid), mainly to people in the low-income bracket, including orphans and widows.

Agricultural Co-operative

This is one of 46 agricultural co-operatives that exist in the New Valley governorate, which was established in 1962. The Agricultural Co-operative in Rashda plays an important role in providing farmers with production inputs such as fertilizers, seeds, and chemicals. In addition, it holds symposiums to enable farmers to acquire the necessary knowledge and skills in new agricultural methods intended to increase agricultural production

Rashda had 719 member farmers in 2006, of whom five were elected as members of the executive board. The board is composed of a director, a secretary, a treasurer, and two ordinary members. It also has a supervisor, who is a permanent officer.

The co-operative's major role is to provide seeds and fertilizers to its members at subsidized prices and to provide agricultural machinery either for rent or for sale. The price of seeds is no cheaper than the market price. However, the farmers prefer to buy from the co-operative, because the quality is said to be better.

The co-operative also provides education and technical assistance for farmers for agriculture, and credit to the members for the production of dates, wheat, and broad beans at an 8% annual interest rate. The co-operative markets major crops for the co-operative's members, such as dates and wheat. The co-operative assists the members to sell dates at market prices to the date factory in Mut city. Through this date factory, dates in Rashda are sold within Egypt or exported to the United States or East Asian countries.

The Youth Centre

A youth centre was originally created as an educational organization in a village that sought to create an athletic generation of physically and mentally healthy villagers characterized by a desire to build and promote their society and to help youths to exploit their leisure time in a healthy and social manner. The Youth Centre was established in Rashda in 1976 and is open to anybody above the age of 18. Each Youth Centre has three playgrounds.

In Rashda, the Youth Centre has around 2,000 to 2,500 members, who pay a small annual fee. It is managed by an executive board elected by its members and manages activities such as football matches and recreational tours.

Education

Table 25 shows the school facilities in Rashda in 2014. The first primary school was established in 1962. Besides Ain Abu Othman, which is a small settlement, the mother village and satellite villages of Rashda and Duhus each have primary schools. Preparatory and secondary schools are only found in the mother village of Rashda. Teachers in kindergarten, primary, and preparatory schools are mainly women, although this is not true of secondary schools.

Besides ordinary public schools, religious schools (Azhar schools attached to Al-Azhar institution) exist to provide Islamic education. One is at the

TABLE 25 *Schools in Rashda (2014)*

		Schools	Classes	Students		Teachers	
				Male	Female	Male	Female
Primary	Rashda	3	23	303	243	24	67
	Duhus	1	6	21	22	16	1
Preparatory	Rashda	3	14	164	144	25	40
Secondary	Rashda	1	5	60	78	17	18
Primary Azhari	Rashda	1	6	45	43	2	10
Preparatory Azhari	Rashda	1	3	16	13	7	3

Source: Local Unit of Rashda (2014).

primary level with six classes, 89 pupils, and 12 teachers (10 of whom are female teachers) as of 2014. Another is a preparatory school with three classes and 29 students, with seven male and three female teachers.

For younger children, there is one kindergarten managed by the Social Development Association (*jama'iyat al-tanmiya al-ijtima'iya*), one of the active associations in Rashda.

Health

Health facilities are offered by the Health Unit. The Health Unit is the basis of services provided by the Ministry of Health. In Rashda, there is one health unit with a family planning centre and a maternal and infant care service as of 2014. There was one pharmacy, one dentist, two technicians, 12 nurses, and one ambulance in 2014. There is also one government pharmacy and one private pharmacy.

The main services provided by the Health Unit are conducting national campaigns against tetanus, cerebrospinal meningitis, and paralysis, as well as providing vaccinations for students, improving the quality of potable water through regular analyses of water samples, family planning, monitoring the health of the local environment, and registering births and deaths.

Religious Facilities

In Rashda, there is no formal place for village meetings such as a *duwwar* or *dar al-munasabat* (gathering place) as in villages in Upper Egypt. The everyday life in Rashda seems to be arranged around mosques. There are 10 government mosques and two private mosques in Rashda. A government mosque is one registered with the Ministry of Waqf, and it has a permanent *imam* (preacher). Some mosques are attached to charitable associations and have public facilities such as gathering places (*dar al-munasabat*) attached to them.

Services Provided by the Villagers

The above-mentioned organizations that provide public facilities and services are governmental organizations. Besides these organizations, various bodies are committed to providing public services in the village, such as local associations or groups of villagers organized by the villagers themselves (Table 26).

Every association (*jam'iya*) must be registered with the Ministry of Social Affairs. There are five, and all are of a charitable nature. Four of these are the Social Development Association, the Rashda Branch of the al-Sharaiya Association, the Association of Umar Bin al-Khattab Mosque, and the Asso-

TABLE 26 *Various bodies committed to public services in Rashda*

Type of organization	**Local Unit**	**Associations**	**Zakat Committee**	**Between villagers**
Legal status	**Governmental**	**Associations registered with Ministry of Social Affairs**	**Branch of associations**	***Musharaka***
Type of activities	Provide basic services (education, health, social assistance, subsidized food, etc.)	Financial and in-kind assistance to poors	Collect zakat	Contribute to building public facilities
	Provide access to water and sewerage	Financial assistance to female household heads and widows	Provide medical treatment	Gathering place
	Provide services for agriculture (wells, canals, technical education, agricultural credit)	Micro credit	Provide support for poor households to continue education	Financial aid to the needy
	Provide administrative services (register land, issue ID cards, etc.)	Financial assistance to orphans	Provide assistance for funerals and burials	
		Provide marriage support for orphans		
		Offer Coran recitals and other religious education		

Source: Information gathered in February 2015.

ciation of Abu Bakr al-Sadiq Mosque. These associations play a role in providing services, especially to the needy.

A Zakat Committee is a branch of the association that aims to collect *zakat* (alms) and profits for *waqf* (religious endowment) from the villagers, and to distribute financial assistance to the poor in the village. There are two such committees in Rashda. One is the Zakat Committee belonging to the Association of Abu Bakr Mosque.

This association itself is of the Khalwati Sufi order (*tariqa al-khalwatiya*), but its activity is not limited to the *sufi* membership. The money is collected at the time of *zakat al-fitr* (charity given to the poor at the end of Ramadan month), or at time of harvesting for *zakat al-mal* (alms to be paid when the contributions reach a certain amount (*nisab*)), or from water *waqf* from two irrigated date palm districts attached to local wells.

The committee also earns money by selling cattle and sheep hides left over from *'Id al-adha* (the Feast of Immolation) to merchants from Cairo. The profit earned from selling these in 2014 amounted to around 25,000 Egyptian Pounds, which is a good income source for financial assistance to the needy. The committee distributes the money to the households provided that they are poor or have started a project, or are travellers.

Musharaka (*Co-operation between Villagers*) *for Public Utilities*

Another type of service is that organized between the villagers, which is called *musharaka*, meaning co-operation. *Musharaka* is of two types. One is co-operation of a public nature, such as construction of mosques, or paving the road.

In Rashda, all the private mosques (*masjid ahli*) were established by co-operation between villagers. Unlike the government mosques, financed and constructed by the government, the private mosques were financed and built by the villagers themselves. If villagers want to construct a mosque, they submit an application to build one with the Ministry of Waqf and Security.

After obtaining a license, they collect money from the villagers and build the mosque themselves. Usually, the mosques are planned and financed by villagers living in the same district. In these private mosques, an *imam* (prayer leader) belonging to the Ministry of Waqf comes every Friday for prayers, in contrast to the government mosques, which have a permanent *imam* and an employee responsible for management. In addition to the *imam*, all management is undertaken by villagers in private mosques. Cleaning and the *adhan* (call to prayer) is performed by villagers living in the same area by rotation.

At the time of writing in February 2015, a mosque named al-Taqwa mosque is under construction. Planning started in 2012, and a license was obtained from the Ministry of Waqf for its construction. The area, called Nazla Bahariya, was originally a *birka* (drainage pond), so there is still some vacant land for the mosque.

Around 50 villagers in the same neighbourhood (Nazla Bahariya) participate in funding the construction, with each household contributing 20 to 50 Egyptian pounds according to their financial circumstances. Some villagers donate a large amount of money, such as A, who works in UAE and donated 90 million Egyptian Pounds. The villagers also participate in the construction when they have time, in the night-time after work, or on the weekend (Friday and Saturday). At the time of writing, the ground floor is finished with a total investment of 80 million Egyptian Pounds. The first floor, which is planned to be a women's prayer room, office, and library, is under construction.

Musharaka (*Co-operation between the Villagers*) *of Reciprocal Nature*

Another type of *musharaka* is co-operation of a reciprocal nature, such as in the construction of houses, wedding ceremonies, funerals, and harvesting. In general, such reciprocal co-operation is organized among relatives for wedding ceremonies or funerals, and in other cases such as harvesting shared among the villagers regardless of kinship relationships. The following cases are examples.

Case 1 Harvesting of Wheat (May 2007)

Mr. S, the owner of a plot in the district irrigated by the Investment Well,[7] called 10 people to harvest wheat at his plot in May 2007. They had no parental relationship with the owner. Five men from Rashda village and two men who are *ʿarab* (Bedouins) living near the village participated in return for a daily wage, and three men from the village of Mushiya participated in reciprocity without monetary compensation. They worked for seven days until the harvest was finished.

Case 2 Harvesting of Dates (September 2007)

Mr. N, the owner of a plot in the district irrigated by the Investment Well, called 15 men to harvest the dates from palm trees on his plot. They worked for two days per week for four to five weeks in September to collect the dates from 8 am to 12 pm, and from 5 pm to 8 pm. 12 men worked for a wage of 70 Egyptian pounds per day, and three men worked for in-kind compensation. The latter men are relatives of the owner, and two received 5 kg of sugar and one received

7 As for the Investment Well and the Well No. 3 Irrigation District, see Chapters 10 and 12 respectively.

a microwave as compensation. Seven men are from different villages; two are from the village of Rashda, one is from Jadida, one is from Mushiya, and eight are from Abu Uthman, where the owner lives. Three men who participated for in-kind compensation are also from Abu Uthman but have no parental relationship with the owner of the plot.

Case 3 Wedding Ceremony (August 2007)

Mrs. F called three men and five women to prepare for her son's wedding ceremony. They are from same family; one came from Cairo to help her. The wedding ceremony took one day, but it required more than a week for preparation, including preparation of food for 500 guests such as washing rice, making bread, and cutting vegetables. It takes two days to prepare the tent for the ceremony, equipping it with items such as kitchen utensils and chairs, and another day before the ceremony. After the ceremony, all these items must be cleaned. They worked in return for the same work to be performed for their children's wedding ceremonies.

PHOTO 23 *View of the Rahman Mosque Street and Izba Shamrukh area (left) in the "new" Rashda*

PHOTO 24
"Commercial Street" (Tijari Street)

PHOTO 25
Youth Centre with the "old" Rashda behind it

PHOTO 26
Social Unit in Rashda

CHAPTER 9

Family Structure in Rashda

In Chapter 3, the structures of household and family in villages in Dakhla Oasis, including Rashda, were discussed from a historical perspective This chapter deals with Rashda exclusively and examines the features of household and family in Rashda through a comparative study of the years 1861 and 2005. The source material for 1861 is the population register of Rashda for that year, and for 2005 is the results of our household surveys conducted in Rashda that year. We first analyse the household structure and cycle in 1861, and then the household and family structure in 2005, and conclude by examining continuity and change in the family structure in Rashda over 150 years.

Household Structure in 1861

Among the documents that we collected during field research in Rashda, we found a copy of the population register of Rashda for 1861 (*Daftar ta'dad nufus hissat al-shaykh ahmad al-hajj muhammad 'abd al-hafiz*) dated 9 Jumada II 1278. The copy, which consists of four papers, is not a complete one; it does not cover all the households but only some. However, it contains enough data to reveal the characteristics of the household structure in Rashda in the middle of the 19th century.

Classification of Household

As mentioned in previous chapters, the family or household are usually called aila (*'a'ila*), *usra* and sometimes *bait* in Egypt. *Bait* means a house. However, in the documents on population in the 19th century, the word *manzil* is used as a unit by which to register village inhabitants. The population register of Rashda for 1861 is such a case. The Arabic word *manzil* literally means "house", the space in which the family lives, like "household" in English.

According to a document dated 28 Jumada II 1279 AH (1862 AD) concerning an incident of arson, which was examined in detail in Chapter 5, 92 *usar* (sing. *usra*) resided in the "old" Rashda on the sand-hill. *Usra* in this document probably signifies "household". The copy of the 1861 population register of Rashda, on which the discussion here is based, contains 50 *manzil*. This means that the copy covers about 55% of the total number of households in those days, if the word *manzil* in the register signifies *usra* in the document on the arson.

 | DOI 10.1163/9789004317390_011

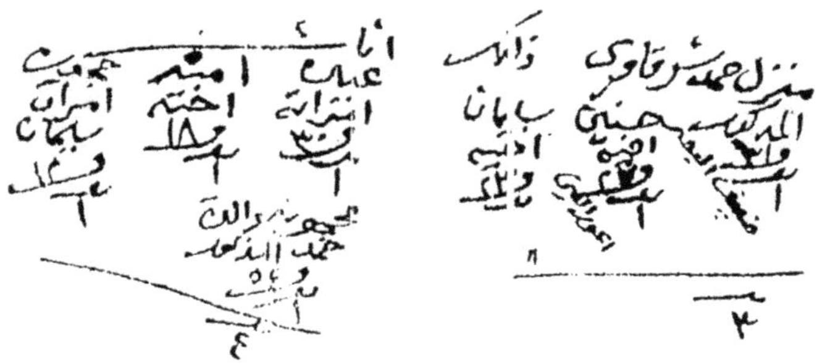

FIGURE 3 *Part of the population register (1861). The right item is an entry for men, and the left item is an entry for women.*

The inhabitants of Rashda are registered by *manzil* in the population register. Each *manzil*, or household, is named after its head, with his age. All the members of the household, relative or non-relative members (329 in total, 173 men and 156 women) are listed with their ages under the name of the household head, ordered by sex and then by relationship with the household head—son, brother, wife, daughter, mother and so on—(Figure 3). Some of them are referred to by their physical characteristics such as "weak-eyed", "squint-eyed", "one-eyed", and "damaged leg".

In Arab society, there appear to be two principles of household formation. The first is the vertical relationship of paternity (the father–son relationship), and the second is the horizontal relationship of brotherhood (fraternal relationship).[1] These two principles determine household composition and size. Therefore, we can categorize the members, relative and non-relative members of a household according to these two principles.

Category 1 (Father–son relationship). This category is paternity based and has three types.

- Type 1 with a married son
- Type 2 with unmarried and married sons
- Type 3 with an unmarried son

Category 2 (Fraternal relationship). This category is fraternity based and has three types.

- Type 4 with an married brother
- Type 5 with unmarried and married brothers
- Type 6 with an unmarried brother

1 This is called ʿ*asabiya* in Arabic.

Category 3 (Father–son and fraternal relationship). This is category is mixed—paternity and fraternity based—and has three types.

- Type 7 with unmarried sons and married brothers
- Type 8 with unmarried sons, married and unmarried brothers
- Type 9 with unmarried sons and unmarried brothers

Category 4 (Incomplete household). This category represents incomplete and unstable households that have no son or brother, meaning that the paternal or fraternal principle of household formation is absent.

- Type 10 without son or brother

This categorization apparently indicates that the households in the 1861 population register of Rashda were formed according to two principles of household formation: paternity and fraternity.

Category 1, which is based on paternity, and Category 2, which is based on fraternity, contain 17 and 13 households respectively, and the household in which both sons and brothers of the head reside is only one of 30 in these two categories.

In addition, all the sons who reside with the brothers of the household head in the 11 households of Category 3, which is based on a mixture of paternity and fraternity, are unmarried. We could deduce from these two facts that paternity and fraternity are two independent principles of the formation of households, although they are not mutually exclusive.

Composition of Households

Table 27 shows the composition of households according to Categories 1–4 and their Types1–10

Households are distributed almost equally among the categories and types, except for Type 8, which has only one household. The percentages of households in each category are 34% (Category 1), 26% (Category 2), 22% (Category 3), and 18% (Category 4). This appears to indicate that there was no priority between paternity and fraternity in the formation of households[2].

The average size of households was 6.6 people. The category that had the largest average size was Category 3 (10.2 people), especially Type 7 (13 people). The category with the smallest average size was, as easily imagined, Category 4

2 According to Cuno (2015): 61, the proportion of household types in the four villages in Lower Egypt in 1847–48 is as follow. Solitaries and no-family households (17.2 %), Households with one conjugal couple (65.1 %), Joint family households (16.7 %), Incompletely classifiable (0.8 %).

TABLE 27 *Compositions and other characteristics of the households in 1861 according to the two principles*

	Category 1 (father–son relationship)				Category 2 (fraternal relationship)				Category 3 (father–son and brother relationship)				Category 4 (incomplete households)	Total
	Type 1	Type 2	Type 3	Total	Type 4	Type 5	Type 6	Total	Type 7	Type 8	Type 9	Total	Type 10	
Number of households	3	4	10	17	4	3	6	13	6	1	4	11	9	50
(%)	6%	8%	20%	34%	8%	6%	12%	26%	12%	2%	8%	22%	18%	100%
Number of household members	26	33	49	108	36	26	26	88	78	10	24	112	21	329
(%)	8%	10%	14%	33%	11%	8%	8%	27%	24%	3%	7%	34%	6%	100%
Average household size	8.7	8.3	4.9	6.4	9	8.7	4.3	6.8	13	10	6	10.2	2.4	6.6
Average age of household heads	69.0	54.0	43.0	51.1	44.5	34.7	24.4	31.5	46.7	40.0	32.0	40.7	23.0	40.1
Number of labor active members	5	16	17	38	8	13	7	28	23	4	9	36	5	107
Number of dependents per labor active member	5.2	2.1	2.9	2.8	4.5	2	3.7	3.1	3.4	2.5	2.7	3.1	4.2	4.2
Households including mother of the heads	0	0	0	0	1	3	4	8	1	1	2	4	4	16
Households including sister of the heads	0	0	1	1	0	1	4	5	1	0	0	1	3	10
Poligamous households	0	0	0	0	1	0	1	2	2	0	1	3	0	5
Plural households	2	1	0	3	2	1	0	3	2	0	0	2	0	8
Households including three generations	2	1	0	3	1	0	0	1	0	0	0	0	0	4

Notes: 1. A plural household is one in which married sons or brothers reside with the household head; 2. A polygamous household is one with more than one wife; 3. Thirteen of 16 households with a mother are with her eldest son. This means that the mother preferred to live with her eldest son, if she had more than one. Three other households are with her younger sons. A household with a mother relates to the principle of fraternity in the formation of the household, because the mother is a kind of "rivet" of the family that makes it possible for brothers to live with their mother; 4. Households with sisters: Nine of 10 households of this kind are also households with a mother. Three households with

(2.3 people). Category 1, based on paternity, and Category 2, based on fraternity, had almost the same average size: 6.4 and 6.8 people respectively.

As mentioned above, the population register of Rashda in 1861 had 329 household members in total, among whom 173 were male (52.6%) and 156 were female (47.4%). All of them were relatives.[3] As for age composition, the oldest man was aged 76 years, and the oldest woman was aged 62 years. The female rate was higher than the male rate in Categories 2 and 4. The higher proportion of women may be related to the instability of these two categories compared with Categories 1 and 3.

Stability of Household

As discussed above, it seems that there was little difference in preferences for the four categories of household formation, because they were of almost equal proportions. However, a household distribution of almost equal proportions could be observed in the year 1861 by chance. We can speculate that this is because of differences in the degree of instability among the four categories.

A household is an arena where the family is reproduced, and a stable household is necessary for the reproduction of the family. Here, the stability of the household is defined by the available labour force that can maintain the female and juvenile members of the household. The reason that paternity and fraternity could be the principles on which households are formed is that these two configurations make it possible to form a stable household.

We take the proportion of labour force and dependent members among all members of a household and indicators of the cycle of a household, which are related to the age of the head and household members, to deduce stability.

The average proportion of labour force members (male members aged from 11 to 60) is 32.5%, and the average total number is 2.1 people. By category, the average proportion and the average number for the four categories are as follows: 35.2% and 2.3 people for Category 1, 31.8% and 2.2 people for Category 2, 32.1% and 3.3 people for Category 3, and 23.8% and 0.6 people for Category 3.

The average number of dependent members per labour force member is 4.3 people, and the average number for Categories 1, 2, 3, and 4 is 2.8 people, 3.1 people, 3.1 people, and 4.2 people respectively. The instability of Category 4 households is apparent because of the small number of labour force members and the heavy burden on them for the maintenance of the household.

3 It can be observed in some documents in the middle of 19th Century that non-relative members such as slaves resided with family members in households, although they were few. See, for example, Kato (1993) and Cuno (1999).

Household Cycles

The indicators related to the age structure of household members, especially the age of the household head, are important for understanding the household cycle, because they influence the number of labour force members by transforming the size and composition of a household through generational change. Here, we take the age of the household head as the indicator concerning the household cycle.[4]

Generational change is an essential key concept for understanding the process of the birth and lapse of households in Rashda, based on data from the population register for 1861, because there was no household—either paternal or fraternal—with more than three generations living together.

Table 28 shows the distribution of household heads' ages by category.[5] The average age of a household head is 40.1 overall, 51.1 in Category 1, 31.5 in Category 2, 40.7 in Category 3, and 23 in Category 4. The lowest average age in Category 4 is no doubt because the households in Category 4 are considered to be an incomplete type of household.

Regarding other types of household, the highest average age is found in Category 1 and the lowest in Category 2, and the average age in Category 3 is between the two. We suppose that this corresponds to the degree of instability of household.

Category 1 appears to have constant stability, because relationships based on paternity are more stable than fraternal structures in the formation of households. However, they are fragile by nature, because such relationships do not continue beyond the father–son relationship, as shown by the scarcity of households with people from more than three generations. Therefore, the households in Category 3 tend to be of the Type 3 subcategory, which is households with unmarried sons; in other words, nuclear families, after the death of a household head.

Category 2 is based on the principle of fraternity, of which the most remarkable examples are residences without sons, married or unmarried. This category seems to be preferred for obtaining labour, when the household head has no son. Therefore, Category 2 does not appear to be as stable as Category 1 for the formation of a household, although fraternity is widely believed to be at the core of social relationships in Arab society.

---

4 Besides the age of the household head, we can calculate the age difference of a couple, and the age of a woman's first marriage and first birth from indicators in the population register. For these details, see Kato (2010).

5 For more detailed data on the age composition of household heads, married sons, married brothers, and mothers living with household heads, see Kato (2010).

TABLE 28 *Distribution of household categories by household head in 1861 (unit: household)*

Category	I Paternity		II Fraternity		III Combination of paternity & fraternity		IV Incomplete "manzil"	Total
Type	1. With married & unmarried sons	2. With unmarried sons	3. With married & unmarried brothers	4. With unmarried brothers	5. With unmarried sons & married brothers	6. With unmarried sons & unmarried brothers	7. Without sons or brothers	
Less than 15	17	13	18	8	40	11	4	111
15-24	18	4	13	6	15	8	5	69
25-34	6	6	13	4	11	5	2	47
35-44	1	4	9	0	12	1	3	30
45-59	10	7	5	4	9	2	1	38
60 & older	3	0	0	1	1	0	1	6
Unknown	4	11	4	3	0	2	5	29
Total	59	45	62	26	88	29	21	330

Source: Kato 2010: 93.

The instability of Category 2 households is obvious in the age difference between the household head and his brothers living with him, and in the fact that many households of this category are those with mothers and sisters. In this case, the role of the mother is decisive for the household to continue with the mother when the household head—her husband—dies.

Households in Category 3 are based on both formation principles—paternity (father–son relationship) and fraternity (relationships between brothers)—so households tend be on a large scale. In fact, the average size of households in Category 3 is considerably larger than that in Categories 1 and 2. Judged by the scale of households, those in Category 3 seem to be stable. However, this is misleading, if one recalls that there are no households with married sons. If an unmarried son were to live in a household of this category, he would leave and form a new household.

As mentioned above, Category 4 has incomplete and unstable households, because it has no principle on which the formation of the household is based, either paternal or fraternal. In fact, the average household is much smaller, and the head is considerably younger, in this category than those in other categories. In addition, households in this category have many dependent members—mothers and sisters—and seven of the nine households are without a household head or his wife. Therefore, because adoption is illegal under Islamic Law, households in this category could collapse unless the household head marries and has a son.

We summarize the discussion to this point on the characteristics of the household cycle in Rashda in the following four points.

1. Households are formed according to one of the two principles of paternity (father–son relationship) or fraternity (relationship between brothers). These two principles are independent but not mutually exclusive. They were compatible at the time of the 1861 population survey, as seen in the almost equal proportions of the four categories of household discussed above.
2. Households based on paternity are relatively stable, because they rely on the basic principle of reproduction of households across generations. However, households with paternity as their foundational principle do not continue beyond two generations. Generally, a son leaves his father's household and forms his own household when he marries. As a result, households based on paternity tend to reduce in size and to become households with unmarried sons; that is, nuclear families.
3. In contrast, households based on fraternity are relatively unstable, because this pattern is preferred when the household head has no son.

The purpose of this arrangement is to obtain labour and to maintain the family. The instability of households based on fraternity is typically reflected in the large number of households with mothers and sisters of the household head. Mothers play a decisive role in the formation of households of this kind, because they prevent their sons from leaving when the head of household—their father—dies. She is the mainstay of the family and has joint possession of family land, which is discussed in Chapter 14. However, when she dies, the fraternal community of her sons is broken.

4. It is certain that households based on paternity are relatively stable, whereas households based on fraternity are relatively unstable. Nevertheless, both types tend to reduce in size between generations. This ultimately results in many small households, some of which are incomplete and unstable, like those in Category 4 without sons or brothers. The average life expectancy of women was short in the 19th century, so a household head had to appeal to remarry when his wife died, or to become polygamous to maintain the size of his household.

Household and Family Structure in Rashda in 2005

Household Structure in Rashda

The structures of households and families in Rashda today have already been discussed and compared with those of another 18 villages in Chapter 3, based on household survey data for 2005. However, it would be useful to summarize some of the main points from this discussion.

Households in Rashda are overwhelmingly nuclear family households (parents and children), classified as Type 3 of Category 1 as above, as are those in other villages in the oasis region.

Rashda belongs to the village category, which rarely includes households that have members based on both relationships that may influence the living arrangements: paternity and fraternity. They are predominantly composed of couples and their children, indicating nuclear family households.

As for the age structure of households, the percentage of the population in the labour force (15–64) in Rashda is a little higher than that of Egypt overall: the percentages for those aged under 15, 15–64, and over 65 are 26.3%, 69.0%, and 4.7%, in comparison with 33.7%, 61.9%, and 4.4% for Egypt overall, respectively. The structure of the labour force reflects the age and sex of household heads.

As for marriage patterns, 63.1% of household heads in Rashda marry villagers with whom they have no kinship tie. Thus, villagers in Rashda are more likely to marry outside their families than are those in the Nile Valley or Kharga oasis.

Family Structure in Rashda

In terms of family size, it should be emphasized that there is a sharp contrast between villages in Upper Egypt and those in Lower Egypt and the oasis region, including Rashda. In Abu Senita in Lower Egypt and Rashda in the oasis region, most families are composed of a small number of households: 75.2% and 54.8% of the respective households belong to families with fewer than 30 households, while people in Homa and Awlad Sheykh in Upper Egypt have larger families: 54.0% of the households in Homa belong to families with 40 or more households, and 94.2% of the households in Awlad Sheykh belong to families with more than 50 households.

Differences are also found in the percentage of households for which information on family members is unavailable, or those that are originally from outside the village. In Rashda, such households are rare, and there are many families that are perceived by the villagers as originally from outside the village. Thirty-nine of the 72 families are originally from outside the village—31.5% of households belong to these families.

This structure is reflected in the residential pattern of families in villages. In two villages, especially Awlad Sheykh, residential space appears to be divided into different family (aila) quarters. In contrast, this spatial arrangement is almost completely absent in Abu Senita (Lower Egypt) and Rashda (the oasis region). In these two places, there is no clear spatial division by family. All the families are small, and they are dispersed over the residential space.

The most impressive deduction from the comparative study of households and families in Chapter 3 is the similarity between Abu Senita in Lower Egypt and Rashda in the oasis region. These two villages are quite different, spatially and historically.

Abu Senita is an "old" village, and its name dates back to the Middle Ages. It is located in the centre of the Delta region, where intensive agriculture based on the water of the Nile has developed from former times.

In contrast, Rashda is a "new" village formed by migrants in the 19th century. It is located in the oasis region of the Western Desert far from the Nile, where the agriculture is dependent on a limited volume of underground water.

It is very interesting to ask why such a remarkable similarity in household and family structures can be observed between the two villages. The key to answering this question is in the discussion in Chapter 4 on the regional categories of socio-economic circumstances.

Chapter 4 clarified the similarity between Lower Egypt and the oasis region, in terms of socio-economic circumstances, such as household income, employment, and educational level. In addition, one of the most decisive factors is the development policy of the state, which is discussed in greater detail in Chapter 10.

Families in Rashda across 150 Years

Comparison of Household Composition

We have conducted a historical analysis of the transformation of households and families in Rashda between 1861 and 2005. The most striking fact in this comparison is that household composition in 1861 and 2005 differs greatly, as shown in Table 29.

In 1861, fraternity was compatible with paternity in the formation of households, as the very similar proportions of the four categories of households show. In contrast, few households in 2005 included brothers.

Moreover, few households included married sons. Households with unmarried sons (type 2) comprised 68.7% of households in 2005, the largest proportion, in contrast to 18.0% in 1861. Households in Rashda in 2005 were overwhelmingly composed of nuclear families (parents and children). These comprised 74.7% of the total. This indicates a decrease in the average size of households from 6.6 in 1861 to 4.2 in 2005.

TABLE 29 *Household composition in 1861 and 2005*

Category of households	Type of households	1861	2005
Paternity	1. With married & unmarried sons	7	7
	2. With unmarried sons	9	378
Fraternity	3. With married & unmarried brothers	7	0
	4. With unmarried brothers	6	1
Combination of paternity & fraternity	5. With unmarried sons & married brothers	7	1
	6. With unmarried sons & unmarried brothers	5	4
Incomplete "manzil"	7. Without sons or brothers	9	159
Total	Total	50	550

Source: Kato (2010) and 2005 Rural Household Survey data.

The decrease in household size in Rashda is dramatically reflected in the increase in the number of households classified as "incomplete" (without sons or brothers) in terms of the classification above. Such households comprised only 18% of the total in 1861 but 28.9% in 2005. These are mainly households of couples or couples and female children.

Why the Difference in Household Composition between the Two Years?

Does the above difference in household composition between 1861 and 2005 imply that family systems have changed drastically? The difference could be partially explained by the way in which households are registered. It was not customary to register households by the name of a female household head in 1861, which could have led to the increase in more numbers of "incomplete households" (with no male members) in 2005 than 1861. In 2005, 4.6% of household heads in Rashda were female. Table 30 shows the distributions of household age in 1861 and 2005. It shows younger household heads in 1861, whereas in 2005, the majority of households had heads older than 45.

Nonetheless, this change in registration affairs is not a satisfactory answer to the question above. To answer it, we should examine household composition in detail. Table 31 shows the number of household members (sons, brothers, and other relatives) per 100 households in 1861 and 2005.

As can be seen in Table 31, the numbers of spouses, sons, and daughters per 100 households are same in 1861 and 2005. This means that although household composition seems to differ, the core of a household is a nuclear family (parents and children).

TABLE 30 *Age of the household heads in 1861 and 2005*

	1861	%	2005	%
Less than 15	3	6.0		
15–24	3	6.0	7	1.3
25–34	8	16.0	100	18.2
35–44	10	20.0	148	26.9
45–59	16	32.0	208	37.8
60 & older	4	8.0	87	15.8
Unknown	6	12.0		
Total	50	100.0	550	100.0

Source: Kato (2010) and 2005 Rural Household Survey data.

TABLE 31 *Comparison of household members in 1861 and 2005*

	1862		2005	
	Number	Number per 100 households	Number	Number per 100 households
Head	50	100.0	550	100.0
Head's spouse	41	82.0	518	94.2
Son	60	120.0	659	119.8
Daughter	35	70.0	513	93.3
Brother	37	74.0	10	1.8
Brother's wife	21	42.0	1	0.2
Brother's children	31	62.0		
Son's spouse	10	20.0	18	3.3
Son's children	8	16.0		
Mother	14	28.0	24	4.4
Brother's mother	1	2.0		
Cousin	1	2.0		
Cousin's spouse	1	2.0		
Brother's grandchildren	1	2.0		
Cousin's son	3	6.0		
Sister	14	28.0	4	0.7
Unknown	2	4.0		
Father			4	0.7
Other			4	0.7
Total	330	6.6	2305	4.2

Source: Kato (2010) and 2005 Rural Household Survey data.

A large difference between 1861 and 2005 is in the number of brothers. In fact, the proportion of households that include brothers has decreased from 74% to 1.8% in modern Rashda.

As mentioned above, household size differs between 1861 (6.6 people) and 2005 (4.2 people). This is explained by the decrease in the number of brothers and their spouses/children. This demonstrates the weakness of fraternal relationships as a principle for the formation of households.

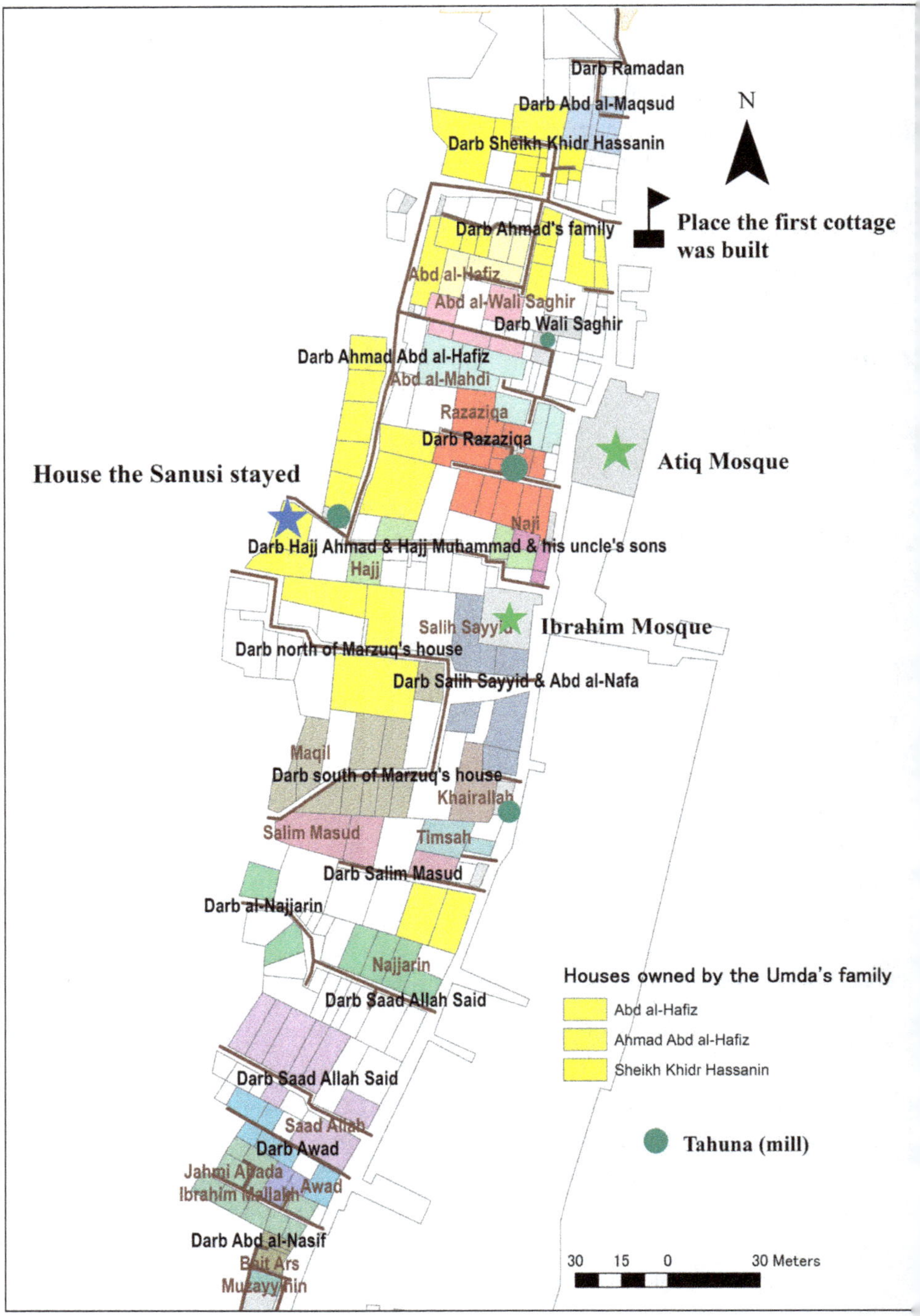

MAP 19 *Location of the quarters* (darb) *and households* (manzil) *by aila in 1861. The source material is a document dated 28 Jumada II 1279* AH *(1861). We examined the owners of the houses in "old" Rashda and placed the relevant information on the map (see Table 14).*

Fraternity as a Safety Valve of Household

The differences in household structure between 1861 and 2005 may appear to represent a change of family system over 150 years. The comparison of the residential patterns of households between 1861 and 2005 are shown in Map 19; Map 20 seems to confirm a change of family system.

Map 19 shows the location of households (*manzil*) in the "old" Rashda in 1861. It shows the location of their residences on the map of the "old" Rashda that we elaborated by the field survey. From this mapping of the 1861 register, it is clear that the households belonging to a family (aila) lived close to each other in one or more quarters (*darb*).

As explained in Chapter 6, the inhabitants of the "old" Rashda moved to the "new" Rashda from the 1960s. Map 20 visualizes the location of households that moved from the "old" to the "new" Rashda by family (aila) name. Households of the same family that had lived near each other in the "old" Rashda tended to disperse or at least not to reside in an obviously collective pattern in the "new" Rashda.

The question of why family households in "new" Rashda do not reside in close proximity is deeply connected with the question of why brothers tended to live together in the "old" but not the "new" Rashda, because the differences in household composition between 1861 and 2005 lie in the number of brothers per household. Our answer to this question is that we must consider it not as a change of family system but as a household arrangement engendered by the household cycle and the socio-economic conditions surrounding the households.

Regarding the household cycle, the following two points are important. The first is the tendency towards decreasing household size discussed above. The second is the role of brothers in household formation. As also discussed above, household heads tend to live with their brothers only when they have no married sons.

This means that a brother is subordinate to a son and that brothers may live together only when there are certain conditions that require cohabitation. Therefore, we consider that fraternity, although it is a principle of household formation, is a relationship that may change over time. In other words, fraternity acts as a safety valve in maintaining the labour force.

That brothers act as a safety valve to maintain the labour force is one thing,[6] but having brothers living with other family members is another. To the question of why the family households do not maintain their proximity in the "new" Rashda, although they had lived collectively in the "old" Rashda, the villagers

6 On the agricultural co-operation among relatives, see Chapter 14.

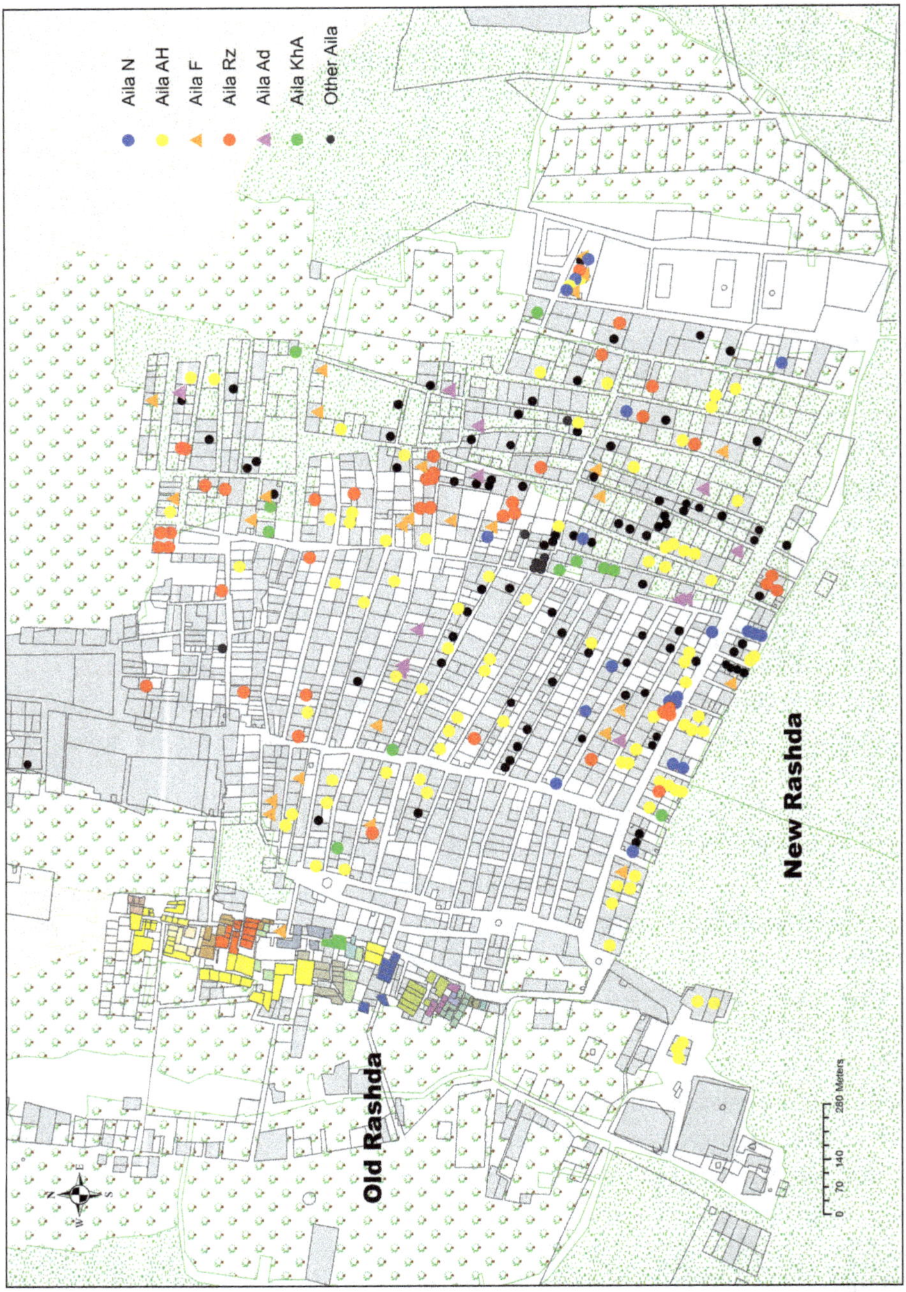

MAP 20 *Location of households that moved from "old" Rashda and settled in "new" Rashda by name of family (aila) (unit: household)*

usually answered, "although the living space in the "old" Rashda was narrow and unhealthy, we were obliged to live side by side".

We also need to examine the changes in socio-economic circumstances at the local and national levels, which are discussed in Chapter 4, and to research the socio-economic functions of families within the village economy, which is discussed in Chapters 3 and again in Chapter 11.

Excursus 3

Family Tree

Social relations in Arab society are principally based on blood relationships. In this context, "blood relationship" formally means the paternal line of family descent, whether real or imaginary. Noble families have family trees showing their line of descent from an honourable ancestor.

Figure 4 is the family tree of the Fuqaha family, which is the family of the *ma'dhun* in Qalamun village. As mentioned in Chapter 5, Qalamun was one of the administrative centres, and the family of the *ma'dhun* insists on its religious dignity and is respected by villagers as an honourable family.

The religious honour of the Fuqaha family is shown in the family tree by the fact that the lineage can be traced back to the Prophet Muhammad. However, it is more interesting in the context of local society that the family tree divides into four branches, three of which are the Qalamun family, the Rashda family, and the Awina family.

As mentioned in Chapter 6, Rashda belonged to Qalamun, and Awina was part of Rashda (see also Excursus 2 on the village cemetery of Rashda). This means that the family tree represents the process of the formation of Rashda village. The following is the translation of the commentary on the family tree.

> May Allah bless our Prophet Muhammad, his family, and his companions, and peace be upon them.
>
> Whatever opinions are held by our eminent *imam* (religious leaders), may all Muslims benefit from them, and may the insurgents be overpowered by them, and may we gain from them both in our present life and in the hereafter, amen, with regard to the below mentioned list of names, whose noble line of descent originates from Abd al-Manaf; namely *Shaykh* Yahia Ibn Mahram, Ibn *Shaykh* al-Salih Yahia Ibn Yahia Ibn Muhammad, son of our late noble master, *Sayyid-na* and *Mawla-na Shaykh al-imam* and great pillar of strength and Muslims' *mufti* (eminent

scholar of Islamic law), the authority, scholar and learned *Hajj* Shams al-Din Muhammad, son of late Shihab al-Din Ahmad, son of late Nasr al-Din Ibn Yusuf, Ibn Muhammad, Ibn Ahmad, Ibn Abu Bakr, Ibn Ismail, Ibn Nasr Allah, Ibn Badr, Ibn Uqayl, Ibn Mashhur, Ibn Abd al-Mansur, Ibn Ahmed, Ibn Abd al-Salam, Ibn Ahmad, Ibn Muhammad, Ibn Isa, Ibn Abd al-Razzak, Ibn Khaled, Ibn al-Jahm, Ibn Khalid, Ibn Abd al-Rahman, Ibn Hashim, Ibn Abd al-Malik, commander of the faithful, Ibn Marwan, Ibn al-Hakim, Ibn al-Aas, Ibn Umaya, Ibn Abd al-Shams, Ibn Abd al-Manaf, also known as al-Mughira, Ibn Qusay, also known as Zayd Ibn Kilab, Ibn Murrah, Ibn Kaab, Ibn Luay, Ibn Ghalib, Ibn Fihr, Ibn Malik, Ibn Anudhar, Ibn Kinana, also known as Quraysh, Ibn Khuzayma, Ibn Mudrika, Ibn Ilyes, Ibn Mudhar, Ibn Mithar, Ibn Muad, Ibn Adnan.

Is this the line of descent of Prophet Muhammad, peace be upon him? Should the lineage of the above-mentioned persons originate from Abd al-Manaf, they would be the proven descendants of Quraysh; if this is so, is it allowed to cast doubts on their lineage? Is anyone entitled to criticize or harm them in any way whatsoever, assuming that they are the noble descendants of Prophet Muhammad? Kindly respond to our enquiry, may Allah, the Giver, the King, reward you and bless you.

It is Allah who grants guidance.

Sayyid-na and *Mawla-na* Muhammad (…) al-Wahhab responded in writing as follows.

Praise be to Allah alone,

This noble lineage meets with that of the Prophet, peace be upon him, at Abd-al-Manaf, as our lord and Prophet Muhammad the Messenger of Allah, may Allah honour him and grant him peace, is the [grand] son of Abd al-Muttalib, who is also known as Shaybat al-Hamd Ibn Hashim.

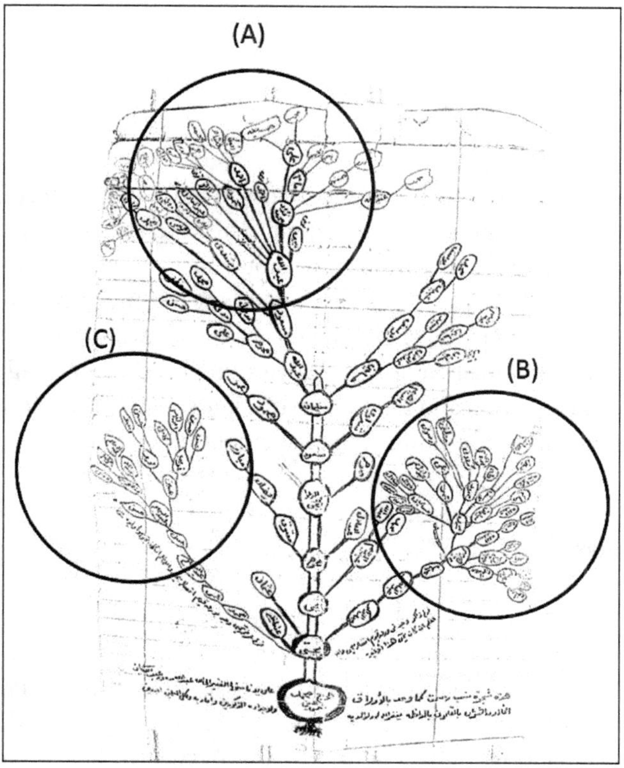

FIGURE 4 *Family tree of the Fuqaha family. The origin of this family tree is Abd al-Manaf, who was a notable of Quraysh tribe in Mekka. The family tree has four branches. Interestingly, the three largest branches are Qalamun (A), Rashda (B) and Awina (C).*

CHAPTER 10

Wells and Irrigation

Besides the effective use of resources, which was discussed in Chapter 2, technical innovation is a key for the continuous development of a society. The most important kind for an oasis village is irrigation technology, because, as we have mentioned repeatedly, life in the oasis region is dependent on underground water drawn from shallow or deep wells. This chapter describes the present situation of wells and springs in Rashda by type (government well, local well, investment well, and surface spring) and the history of their construction, and discusses the system of water distribution and management.

Wells and Springs

Land Register

Livelihood in an oasis depends on underground water drawn from wells. Hence, the oasis has a well-centred society determined by the nature of the well. In other words, the nature of the well determines the pattern of landholding, cultivation, and social relations between the villagers. In fact, the irrigation districts that depend on different types of well have different forms of livelihood.

For irrigation, Rashda mainly depends on ground-water with a minimum potentiometric level of 96 metres, compared with the highest recorded potentiometric surface of shallow wells in Dakhla Oasis, which is approximately 132 metres in the Tenida area on the eastern side. This means that the groundwater moves towards Rashda village from all directions. On the other hand, the potentiometric level of deep wells in Rashda is about 130 metres [Kato et al. 2012: 186].

The importance of wells is expressed in the method of registration of cultivated land in Rashda, and oasis villages in general. The cultivated land of the village is divided into two categories: *zimam* and *kharij al-zimam*. *Zimam* means the land surveyed and recorded in the Survey Department Books and Registers, which is subject to a realty tax on agricultural lands according to Law No. 143 of 1981 concerning desert lands.[1] It is registered as agricultural land belonging to the village.

1 *Zimam* refers to "the limits of the perimeter that comprises urban lands within city or village cordons, as well as cultivated and uncultivated agricultural lands that have been surveyed by

 | DOI 10.1163/9789004317390_012

Kharij al-zimam literally means "outside of *zimam*". The areas outside the *zimam*, which were unsurveyed and unregistered, are owned by the government. The cultivators can usually obtain the right to own the land when they have cultivated it for 10 years after registering it with the land reclamation authority (General Authority for Reconstruction Projects and Agricultural Reclamation (GARPAR)). The government-owned land earmarked for agricultural and reclamation purposes is called *kharij al-zimam.*[2]

In Rashda, the *zimam* refers to the agricultural land that was registered as belonging to the village when the New Valley Project started. The lands registered as *zimam* are those that have been irrigated by local wells from before the beginning of the New Valley Project, or those irrigated by government wells. Thus, the agricultural land is defined by wells. The cultivated land outside the *zimam* is estimated to be 528 *feddan*, compared with the area of Rashda inside the *zimam*, which is 2,500 *feddan*.[3]

Inside the *zimam*, agricultural land is registered by the name of the well that provides, or originally provided, the water for irrigation. As shown in Map 21 on page 194, the *zimam* is composed of several agricultural fields (*mintaqa*) named after the well that provides their water. Hereafter, we call these agricultural fields "districts".

Types of Wells and Springs[4]

In Rashda, wells are classified into five types: government wells (*bir hukumi*), local wells (*bir ahli*), investment wells (*bir istithmari*), surface springs (*'ain sathi*), and Roman springs (*'ain rumani*). Their names imply a distinction between springs (*'ain*) and wells (*bir*). In fact, the natives recognize water sources according to depth; they are called "wells" (*bir*) if the source must be tapped using a drill, or "springs" (*'ain*) if the water bubbles up naturally and

the Egyptian Survey Authority (ESA) and included in the register of agricultural lands and real estate (*kashf al-mukalafa*) maintained by the Real Estate Tax Department (RETD)". Food and Agriculture Organization of the United Nations (FAO) web site: Egypt, National and local institutions enforcing land regulations <http://www.fao.org/gender-landrights-database>.

2 *Kharij al-zimam* originally comes from a decree in 1884, at the time of a cadastral survey in the Nile Valley. It was intended to authorize the sale of uncultivated lands that had not been included in the cadaster but were planned for inclusion with the expansion of cultivated land following the control of the Nile flood (Survey Department 1908: 67).

3 Information obtained in February 2008 from the Agricultural Co-operative in Rashda. The Local Unit estimates the area inside *zimam* to be 2,807 *feddan* and outside *zimam* to be 161 *feddan* (2004).

4 This section is based on Kato et al. (2010).

only needs to be cleared from time to time. However, with the disappearance of artesian wells, *ʿain* is now becoming a historical name.

A more important distinction in the present day is the way in which the government intervenes, as the classification of wells as "government" or "investment" suggests. The form of governmental intervention changes with changes of economic regime. We shall examine types of wells in detail below (Table 32).

There are an estimated 65 wells and springs in Rashda. Among these, the oldest are Roman springs followed by local artesian wells dug in the traditional way (*ahli*). These two types were the only water sources that existed before 1959; the modern history of the wells began with the construction and reclamation of the New Valley in that year.

Roman springs are popularly recognized as dating back to the Roman era, although most of them are assumed to be 200 years old or less. They had a depth ranging from 85 to 100 metres. They have all dried up now.

Local wells (*bir ahli*) are exploited by local cultivators, who dig them in the traditional local way and own and manage them collectively. Most of them were initially dug as artesian wells and had a depth of 85 m or less. In the old days, these traditional wells were dug by local professional diggers called *najjarin.*[5] However, most of these traditional wells have dried up and been replaced by diesel pumps. There is a leader of the well who has a list of the rights holders who have access to the water source. The water source is supposed to last for 30 to 40 years. There were 29 wells of this type in Rashda, of which 13 were operating in 2008.

Government wells are those that have been opened since the 1960s under the Land Reclamation Scheme in the Gamal Abdel Nasser era. They were drilled and dug by the General Authority of Desert Development,[6] and are

---

5 *Nagarrin* or *Najjarin* literally signifies "carpenters".

6 In the 1960s, it was the General Authority of Desert Development that managed the land reclamation projects, including the drilling of government wells. This authority later became the Ministry of Agrarian Reform and Land Reclamation in the 1960s, which separated into two ministries: the Ministry of Agriculture and the Ministry of Land Reclamation. From 1996, they were reintegrated into one ministry: the Ministry of Agriculture and Land Reclamation. "The evolution of the structure of the Ministry of Agriculture" Ministry of Agriculture and Land Reclamation web site <http://www.agr-egypt.gov.eg/En_MinHistory.aspx>, accessed on 27/03/2015. The current organization in charge of the land reclamation is the General Authority for Reconstruction Projects and Agricultural Reclamation (GARPAR), under the Ministry of Agriculture and Land Reclamation (MALR). This organization controls land reclamation projects, and agricultural and uncultivated lands within the arable perimeter, inside the *zimam*, that are not privately owned.

TABLE 32 *Wells and springs in Rashda (2008)*

	Government well	Local well (pump)	Local well (artesian)	Investment well	Surface spring	Roman spring
Depth (m)	1200–1800	500–700	85–120	Below 300 m by law. But actually many are around 500 m	120–150	85–120
Institution of authorization	Ministry of Irrigation	Ministry of Irrigation	By 'putting hand'	General Authority for Investment	Land Reclamation Fund	
Estimated duration of use	50 years	40 years	30 years	30 years	20 years	Unknown
Start year	1961	1907	From before 1800	1995	1996	Unknown
Owner of well	Government		Water use right holders			
Management of water	Irrigation Office & Board		Water use right holders			
Number	7	13	Dried up	10	35	Dried up

Notes: 1. Information is for Rashda village. Budkhulu and Duhus are not included; 2. The 35 surface springs include those that have dried up.
Source: Information collected by informant in 2008.

currently owned and managed by the Irrigation Authority under the Ministry of Water Resources and Irrigation.[7] They generally have a depth of 1,200 metres and are supposed to last for 50 years. There were seven wells of this type in Rashda in 2008.[8] Many of them were drilled at the time of the construction campaign in the New Valley on October 3, 1959. The government wells are drilled as a part of the New Valley land reclamation project scheme, whereby the reclaimed land is given to beneficiaries usually chosen according to the criteria of being impoverished or landless villagers. Each beneficiary receives 4 or 5 *feddan* of land leased from the government and an equal right to water.

Investment wells and surface springs are types of wells that began operating in the 1990s with economic liberalization. Investment wells are exploited by one or several individuals ("investors") under a scheme of the General Authority for Investment (*al-Hay'a al-Amma al-Istithmari*) that began in 1995. This organization is an affiliate of the Ministry of Investment. An individual who wants to start up an agricultural business is allowed to dig a well after obtaining permission from this organization. By law, the depths of wells must be 300 metres, and the area of irrigation of each well is defined as 10 *feddan*. Actual depths and areas are often larger. Many wells have a depth of 500 metres or less and are supposed to last for 20 to 30 years. They are generally co-owned by individuals, and their irrigated land is privately owned after being leased for 10 years from the government. There are 10 wells of this type in Rashda, covering a total area of approximately 950 *feddan* as of 2008.

Surface springs (*'ain sathi*) are popularly recognized as springs that bubble up by themselves, although their depth is generally between 120 and 150 metres. By law, a surface spring's area of irrigation is defined as 10 *feddan*. They are drilled with permission from the Land Reclamation Fund. They are individually or collectively owned and managed by the cultivator(s). Their irrigated land is leased to cultivators by the government, which owns the land. There are about 34 springs of this type. The first was developed in 1996.

Names of the Districts according to Different Types of Wells

The districts irrigated for a long time have the names of Roman springs. For instance, one of the oldest and largest "districts" in Rashda village, with 100 *feddan*, is called Ain al-Rahma. Although Ain al-Rahma has dried up, the district

7 The Ministry of Irrigation changed its name to the Ministry of Public Works and Water Resources in 1987, and in 1999 changed it again to the Ministry of Water Resources and Irrigation. Ministry web site <http://www.mwri.gov.eg/historyofministry.aspx>, accessed on 27/03/2015.

8 In 2013, there were eight government wells.

TABLE 33 *Districts irrigated by local wells in Rashda (2008)*

Districts of date palms			Districts irrigated for crops		
District names	Number of date palms	*feddan*	Name of wells	*feddan*	*qirat*
Bir Sheykh	4,890	23	Bir al-Sheykh	61	19
Ain al-Balad	5,946	40	Ain al-Musid	96	7
Ain al-Rahma	1,268	100	Bir al-Qibli al-Jadid	67	13
Ain al-Shishlana	9,558	8	Bir al-Qadim	76	15
Bir al-Zanjura	4,253	29	Bir Ain al-Balad	41	18
Bir al-Nakhil	8,878	52	Bir al-Jadid	13	18
Bir al-Qadim	1,015	13	Bir al-Qadim	32	13
Bir 4 al-Rashda	193	12	Ain al-Duma	82	15
Ain al-Duma	784	7	Rushayd	9	16

Source: Information collected by informant in February 2008.

is still called by the name of the well that provided water. The second largest district irrigated by the local well is Bir al-Nakhil (Fig. 10–2). These districts are now irrigated by one or more local wells. Sometimes surface springs are also used to complement the water supply.

The districts irrigated by government wells, on the other hand, are numbered. There are four actual districts irrigated by government wells in Rashda: Well No. 1, with 454 *feddan* and 100 beneficiaries, Well No. 2 with 371 *feddan* and 32 beneficiaries, Well No. 3 with 167 *feddan* and 34 beneficiaries, and Well No. 4 with 63 *feddan* and 150 beneficiaries.[9] Surface springs and investment wells are generally named after the leader of the cultivators who helped to drill them. For example, a surface spring drilled in 2003 is called "*Hajj* Hasan" and is shared by 12 cultivators.

9 Information obtained from the Agricultural Co-operative in Rashda in March 2006.

Development of Wells

Years of Construction of Wells

We collected data and information on approximately 70 wells and springs in Rashda Village. The information collected on these wells includes the name, type of well, location, depth, year of drilling, mode of effusion, quantity of water flow at the time of drilling (cubic metres per day) for government wells, and status of operations (whether they are operational).

Table 34 shows types of well classified by year of construction. Besides local wells and Roman springs, all of the wells in Rashda were drilled after the New Valley Project began. Three government wells were drilled in 1959. These are Well No. 1 in Rashda West, drilled by an Italian company that worked on the New Valley Land Reclamation Project; Well No. 2, and Well No. 3, which was also drilled by the Italian company. All of these wells have a depth of 500 metres.

Many wells were drilled after 1996 on the outskirts of Rashda. During 1970 and 1995, only three new wells were drilled. In contrast, drilling of new wells increased tremendously during 1995 and 2004, with 14 new wells drilled between 1995 and 1999, and 26 between 2000 and 2004. This acceleration of drilling was brought about by the authorization of drilling in 1995 in the forms of investment wells and surface springs in particular. Surface springs grew remarkably in number between 2000 and 2004, with 17 new wells drilled during this four-year period.

The surface springs were a preferred type of well for the villagers, probably because of their low cost. As mentioned above, they are shallow wells of 120 or 150 metres. Therefore, they need only a small drill and a small submersible pump, so they do not require a huge amount of money. According to an informant, the cost of drilling was estimated to be around 30,000 Egyptian Pounds in 2008. Investment wells, which have a depth of around 500 or 700 metres, require approximately 300,000 Egyptian Pounds, an amount 10 times more than that of surface springs.

The first two investment wells were drilled in 1996. These wells were drilled by individuals and have a depth of 500 metres. Investment wells have an image of being for "investors" or businessmen, from the outside. However, in case of Rashda, the "investors" are mostly small "investors" who are native villagers in the middle-income category.

A consequence of the increased drilling from the late 1990s is the lowering of the underground water level. An informant and other farmers, recalling their childhood in the beginning of 1990s, said that the artesian wells had func-

TABLE 34 *Year of construction of wells in Rashda by type of well (2009)*

	Local well		Government well		Investment well	Surface spring			Total
	Stopped	Operating	Stopped	Operating	Operating	Stopped	Operating	Unknown	
Before 1950	8								8
1950–1959	4		3						7
1960–1969				1					1
1970–1995	1	1		1					3
1996–1999		3		2	6	2	1		14
2000–2004		4	1	1	3		17		26
2005–2008		3		2	1		2		8
Total	13	11	4	7	10	2	20	3	70

Source: Information collected by informant.

tioned well up to that period: the water came out so strongly that it was difficult to close the valve, but since then, it had become easy to do so.

Interestingly, it seems that the villagers, including the informant and other farmers, view the problem as a matter of uncertainty of water flow, rather than as a risk of exhausting the overall underground water reserves.[10] If the water flow stops from one well, they drill a new well. Thus, the drilling of wells continues and has accelerated in recent years to ensure a water supply, and this has been accompanied by the change in the form of state control over water sources.

Continuous Drilling of Wells

However, we insist that the continuous drilling of wells is not a new phenomenon (Table 35). The water flow has been a major concern of the villagers for a long time. As mentioned above, there were 13 springs (*ʿain*) that existed in 1883 in Rashda according to a document dated 1937 (Table 21 in Chapter 7): Ain al-Rahma, Ain al-Shishlana, Ain al-Musid, Ain al-Jumaiza, Ain al-Balad, Ain al-Duma, Ain Sahsah, Ain al-Rashda, Ain Shandum, Bir al-Majnun, Bir Hamam wa Mahmud, Bir Qara Birqas, and Bir al-Qadim. At present, all these Roman springs—of which five (Ain al-Balad, Ain al-Duma, Ain Sahsah, Ain al-Rashda, Ain Shandum) already had no water flow in 1937—have dried up.

Ain al-Rahma, believed to be the oldest of the Roman Springs, was replaced by Bir Hamam wa Mahmud when it dried up in the 1930s. However, the water flow of the latter well also decreased. An informant recalls that he swam in the Bir Hamam wa Mahmud in 1980s, but it finally dried up in the 1990s. Then a new well, Bir al-Jadid, was drilled in 1929 to support the water supply but it also declined. In 1985, Well No. 20/4 (or Bir 4 Gharbi meaning Well No. 4 West) was drilled to a depth of 543 metres. However, because of its insufficient water supply, a surface spring (*ʿain sathi*) was dug to a depth of 75 metres in 1994, followed by Well No. 13/4 (or Bir 4 Sharqi meaning Well No. 4 East), which was drilled in 2003. It is a deep well, with a depth of 1,180 metres.

The water supply from Ain al-Balad also stopped in the same period as Ain al-Rahma, because it is marked "extinguished" in the document dated 1937 (Table 21 in Chapter 7). The well was replaced by Bir Ain al-Balad, dug in 1927, but this well also dried up in the 1950s. Therefore, the agricultural field originally irrigated by the water from Ain al-Balad received water from the government Well No. 3, which will be discussed in Part 4, and from 1977, it was also supplied by a new local well of Dulab Bahry, which dried up soon after. Currently, the water is supplied by Well No. 3 and Haramiya Musid, dug in 2004.

10 A similar remark is made by Viscomi (2010: 93).

TABLE 35 *Replacement of Roman springs to new wells*

			Depth (m)	Year of digging	Current status
Ain al-Rahma	Ain al-Rahma	Roman spring	Around 100	Before 1883	Dried up in 1940s
	Bir Hamam wa Mahmud	Local well		Before 1900	Dried up before 2005
	Bir Jadid	Local well		1929	
	Well No.20/4 (Well No. 4 West)	Local well	543	1985	Operating
	Ain Sliman	Surface spring	75	1994	Operating
	Well No. 13/4 (Well No. 4 East)	Local well	1180	2003	Operating
Ain al-Shishlana	Ain al-Shishlan	Roman spring			Dried up in 1990s
	Well No.2 0/4 (Well No. 4 West)	Local well	543	1985	Operating
Ain al-Balad	Ain al-Balad	Roman spring		Before 1883	Dried up before 1937
	Bir Ain al-Balad	Local well		1927	Dried up in 1950s
	Dulab Bahry	Local well		1977	Dried up
	Haramiya Musid	Local well		2004	Operating
Bir al-Majnun	Bir al-Majnun (Bir Qibli Maarif)	Local well	120	Before 1883	Dried up before 1997
	Well No.11 (Qibli Mukhaiem)	Local well	1050	2004	Operating

Source: Information collected in March 2015.

Another case is Ain al-Shishlana, which is a Roman spring that survived until the 1990s. Until the 1980s, the water flow was so abundant that it was used to drive the wheat mill. However, the spring dried up in the 1990s, and it was replaced by Well No. 20/4, which also delivers water to the district of Ain al-Rahma.[11]

It should be noted, however, that deep wells are unstable. The wells must be continuously dug, but not because they are shallow. In fact, many deep government wells were drilled from the late 1990s, because the three wells drilled in 1962 are no longer in operation. They include two wells in the Well No. 1 Irrigation District, one well in the Well No. 2 Irrigation District, and three wells in the Well No. 3 Irrigation District.

Geographically, the drilling of wells took place in the northern part of Rashda village. In fact, the Roman springs, local wells, and surface springs are all located north of the modern village residential area. The location of the wells is because of the main topographic feature of Rashda, which is surrounded by higher plateaus or hills that are part of the northern plateaus surrounding Dakhla Oasis.[12] Many wells are drilled on this northern plateau rim where the elevation is high, at around 120 metres.

Agricultural fields or districts, on the other hand, are located to the south on the plain and flat area, about 80 to 100 metres above sea level. Several canals connect these agricultural fields with the wells, which run north to south. Thus, the characteristic of agricultural fields in Rashda is that the wells and agricultural fields are in separate locations.

There are several reasons for drilling wells far from the agricultural fields, or siting agricultural fields far from the wells. One reason may be that natural springs and free-flowing shallow wells were easily tapped on the northern fringe of hills (Map 22 on page 195). This may be because of the geological conditions or because the ground-water level is at a shallower depth in the border areas.

Another reason is to ensure maximum water flow using gravity. Slowly running down the slope from the hilltop area, the canals that divide into branch canals direct the water flow to different fields. Although the separation of wells and agricultural fields may risk evaporation from the canal, it has been an efficient method for maximizing water flow. Beadnell, visiting Ghenina village in Kharga Oasis noted that "although actual difference of level is so little, amounting to seven or eight metres, the difference of flows from wells of equal depth on either side of area averages fully 100 percent" (Beadnell 1909: 13).

11 The information on irrigation by these springs and wells in the 19th Century is based on a series of historical documents described in Chapters 5 and 7.

12 Hills have names such as *hajara baydh'* (meaning White Stone).

Deep wells are not limited by slope and can be located at sites convenient for transportation or close to the fields. In fact, most of the government wells are located near their irrigation districts. However, because most of the cultivable land on which these irrigation districts developed is located at the ground level where old districts already exist, the water sometimes needs to be shared with the farmers who have the rights to land in the local well district. This complicated situation is described in Chapter 13.

Thus, the defining characteristic of agricultural fields in Rashda is that each depends on several wells. In other words, a well is a water source for several fields. As a result, water drawn from a well must be shared not only between farmers in the same district, but also between districts.

Water Distribution

Methods of Irrigation

Water flows through primary, secondary, and then tertiary canals (*marwa*) into parcels (*khatt*) of cultivated land. In the mouths of some wells, a v-notch weir, a triangular channel section, is installed to measure water discharge. The majority of the main and secondary canals are lined in all types of well. At each canal junction, there is a gate or a distributor (*maqsam*) to control the water flow.

A water gate is used when the water is distributed periodically, whereas a distributor is used mainly for crops or plants that require permanent water flow, such as rice cultivation or palm tree orchards. The distributor for continuous water supply, known as a rice gauge, is the same as that described by King in 1917: "a long board fixed in the bed of the stream and having its upper edge the same number of notches as there are owners of the well" (King 1917: 360).

The water is then distributed to the plots according to the *ghamr* method or flood (border) irrigation. In Rashda, sprinkler irrigation is not used. Drip irrigation is rarely used in districts irrigated by an investment well.

The *ghamr* method, or flood irrigation, is a method whereby the water from the main or branch canal floods into the parcel until the entire field is covered. Its name changes according to the way a parcel is divided. For instance, the *bakiya* (*nahr*) method, or furrow irrigation, is a method whereby the parcels are divided into divisions (*bakiya*) and irrigated by small canals (*qanat taqdim*) inside the parcels. In this *bakiya* method, a parcel is composed of ridges (*mastaba*), and there are water passages or furrows (*nahr* or *khulayja*) between these ridges. It is used to irrigate small basins for various crops. An another method, called *ghamr kulli* or *batti*, distributes the water collectively into a parcel composed of small basins. It is suitable for the cultivation of crops such as rice, sorghum, Egyptian clovers.

The flood irrigation (*ghamr* method) is the most common method used in Rashda and Dakhla Oasis in general. Almost all the farmers employ this method. The following three factors influence the choice of method. The first factor is the soil. Clay soil is preferable because it retains water for a long time. The second factor is capital. The third factor is the volume of water. The flood method has some advantages in this respect. It is economical, because this method washes and adjusts the soil if it is saline, and it is suitable for grain crops and alfalfa. It is also easy to use if there is abundant water.

Water Distribution

Like other villages in the New Valley governorate, water management practices differ by type of well. In the districts irrigated by government wells, the government has dug deep wells (government wells) where ground-water is available and accessible. Irrigated areas were allocated to beneficiaries with a minimum of 5 *feddan* each. The water quota is theoretically determined as 22 cubic metres per day per *feddan* during the period of high water requirement (summer), while in winter it is less, when the flow from the well is reduced.

In irrigation areas serviced by pump-operated government wells, water distribution is controlled and is tied to pump working hours. A government-employed local farmer operates the pump's generator and supervises operations throughout each day. As the water supply is controlled by the pump, it can be tailored to match demand; for example, by season or crop requirements.

In the local wells, the water quota is inherited from grandfathers who participated in digging the local wells or the spring. Water allocations and distribution are decided by farmers. The leader is called the chief of water (*ra'is al-ma'*) and is nominated by the elders and those with experience. He keeps a register book (*kashf*), manages the water schedule, and mediates any problems with irrigation.

In the investment wells and surface springs, the water quota is decided according to the contribution that each member made to drilling the well and installing the pump. Like local wells, one farmer is nominated as the leader to manage the water schedule.

Water quotas from the local wells, investment wells and surface springs are not set on a volumetric basis but on a time basis. Water is divided into indefinite time shares, in a given water rotation period. Districts or subdistricts rotate water on different time schedules ranging from once every six or seven days to once every 12, 16, or 18 days. The length of rotation varies according to volume of flow and type of crops.

The unit of time used is the *amila,* which equals one day or night (or from sunrise to sunset, or from sunset to sunrise); each *amila* is divided into *qadam,* which usually number 30. The smallest unit is the *habba*, which equals one-quarter of a *qadam*. Thus, one *amila* equals 30 *qadam* or 120 *habba*. This is so even for the district attached to a government well.

Therefore, the most important concern for farmers is the number of *amila* that they receive. This is decided according to their contribution to drilling the well. It should be noted that it is a concept based on time, but it is a relative concept that can be adopted according to the total definite period of time. Thus, it is used to define the share of each shareholder.

Table 36 is an example of a register (*kashf*) that indicates the share of each owner in the local well in the neighbouring village of Hindawi. For local wells, water distribution is based on a register, which sometimes dates from more than 100 years ago. In the case of this local well, there were originally 15 owners. Shareholder A had six *amila* and eight *qadam*, which equals 188 *qadam*. Suppose that the water schedule is for 12 days, then owner A receives 188 / (45 *amila* × 30 = 1,350 *qadam*) × 720 *qadam* (12 days × 60 *qadam*) = 100 *qadam* = 3 *amila* and 10 *qadam*. In this way, owner A takes the entire flow for one and a half days plus 10 *qadam*. Other owners follow in the same way until 12 days of watering are complete.

As King noted about Dakhla Oasis, the watering schedule becomes much more complicated when there are more shareholders, or when the shareholders have different interests in crop cultivation. In such cases, "the difficulty is overcome by the proprietors selling or exchanging the water with each other" (King 1917: 361).

The above case is relatively simple. However, most water distribution systems are much more complex when there are many shareholders and the district is large, or the supply is divided between subdistricts with different crops. The flow of water passes through water dividers that allocate water to specific fields according to a rotation schedule.

In the districts irrigated by the government wells, water distribution takes a different form. It is conducted on a volumetric basis. Table 37 shows an example of a district irrigated by government wells. In the districts irrigated by government wells, a plot of cultivated land should receive 22 cubic metres of water per *feddan* per day.

This quantity is decided by the government, based on the water requirements according to crop and season. In summer, when evaporation is higher, the quantity of water is raised to 30 cubic metres per day per *feddan*. Therefore, the volume of water that every plot should receive is calculated according to water demand. In principle, when the level of water discharge falls below the water requirement calculated on this basis, the Ministry of Irrigation should drill a new well to maintain the required level of water discharge.

However, the water volume is measured only at the well level, not the plot level. Therefore, there is no exact information about the water supply, which sometimes results in disputes over water distribution.

TABLE 36 *Water rights to Ain Munisa in Jabal Husam (Hindawi)*

Name of owners	*Amila*	*Qadam*
Muhammad and his brothers	6	8
Hasan	2	27
Sayyid	5	1
Shaykh Husayn Ibrahim	2	8
Muhammad	0	23
Mahmud	1	21
Hasan and his cousins	3	2
Madani	1	13
Imbabi and his brother's sons	5	19
Muhammad and his brothers	1	8
Fadil	2	25
Ali and his brothers	2	14
Muhammad and his brothers	1	0
Shaykh Husayn Khalil	1	26
Umran and his partners	6	6
Waqf of *Shaykh* Husam	0	2
Waqf of *Shaykh* Abdullah Juhayni	0	2
Total	45	0

Source: A document possessed by a village notable.
Note: *Shaykh* Husam and *Shaykh* Abdullah Juhayni are the names of saints whose tombs are located in the cemetery in Awina. *Waqf* here means that two lots of two *qadam* of water are sold to the farmers and used for charity .

An annual meeting is held to decide or confirm the watering rotation schedule. In the case of government well districts, it is directed by the irrigation official from the Irrigation Office attached to the Ministry of Irrigation in Mut. In the case of other wells, it is decided by the shareholders. Seasonal repairs and cleaning of the wells and canals are required to maintain the flow.

As described in Chapters 12 and 13, these seasonal maintenance works are usually performed by workers hired by the Irrigation Office in the case of government wells, and collectively by users if the area is relatively small in the case of surface springs or investment wells. In the case of local wells, it is usually performed by one of the farmers. He takes charge of the maintenance, in return for an extra water quota. This system of compensation, jaala, is described in Chapters 7 and 12.

TABLE 37 *Estimated quantity of water distribution in one of the subdistricts of Well No. 3 Irrigation District (winter)*

Block	Plot ID	Area (*feddan*)	Time of irrigation		Estimated volume of irrigated water	
			Hour	Minute	(m^3/day)	(m^3/18days)
1	1	1.1	12	20	24.9	447.5
	2	1.1	12	20	24.9	447.5
	3	1.2	13	40	25.7	463.3
	4	1.2	13	0	25.7	463.3
	5	1.2	13	20	25.5	459.4
2	6	1.0	9	20	22.9	411.8
	7	1.0	8	40	22.4	403.9
	8	1.0	8	0	22.0	396.0
	9	1.1	12	20	24.9	447.5
3	10	1.1	11	40	24.4	439.6
	11	1.1	12	40	25.1	451.4
	12	1.1	11	40	24.4	439.6
	13	1.2	13	20	25.5	459.4
4	14	1.1	12	20	24.9	447.5
	15	1.1	12	0	24.6	443.5
	16	1.1	11	40	24.4	439.6
	17	1.2	14	20	26.2	471.2
	18	1.2	15	0	26.6	479.2
	19	1.2	14	40	26.4	475.2
	20	1.1	12	20	24.9	447.5
5	21	1.2	15	40	27.1	487.1
	22	1.2	15	0	26.6	479.2
	23	1.1	11	40	24.4	439.6
	24	1.1	11	20	24.2	435.6
6	25	1.0	9	20	22.9	411.8
	26	1.1	12	40	25.1	451.4
	27	2.0	16	20	44.2	796.0
	28	2.0	16	40	44.4	799.9
	29	1.2	13	20	25.5	459.4
	30	1.2	13	20	25.5	459.4
	31	1.2	14	0	26.0	467.3
	32	1.1	12	20	24.9	447.5
	33	1.2	13	0	25.3	455.4
	34	1.1	12	40	25.1	451.4
Total	595	40.3	419	780	887.5	
			18 days			

Note: For information on Well No. 3 Irrigation District, see Part 4.
Source: Information collected by informant in 2008.

MAP 21 *Boundaries of cultivated lands in Rashda. Source*: 1/2500 scale map (1985).

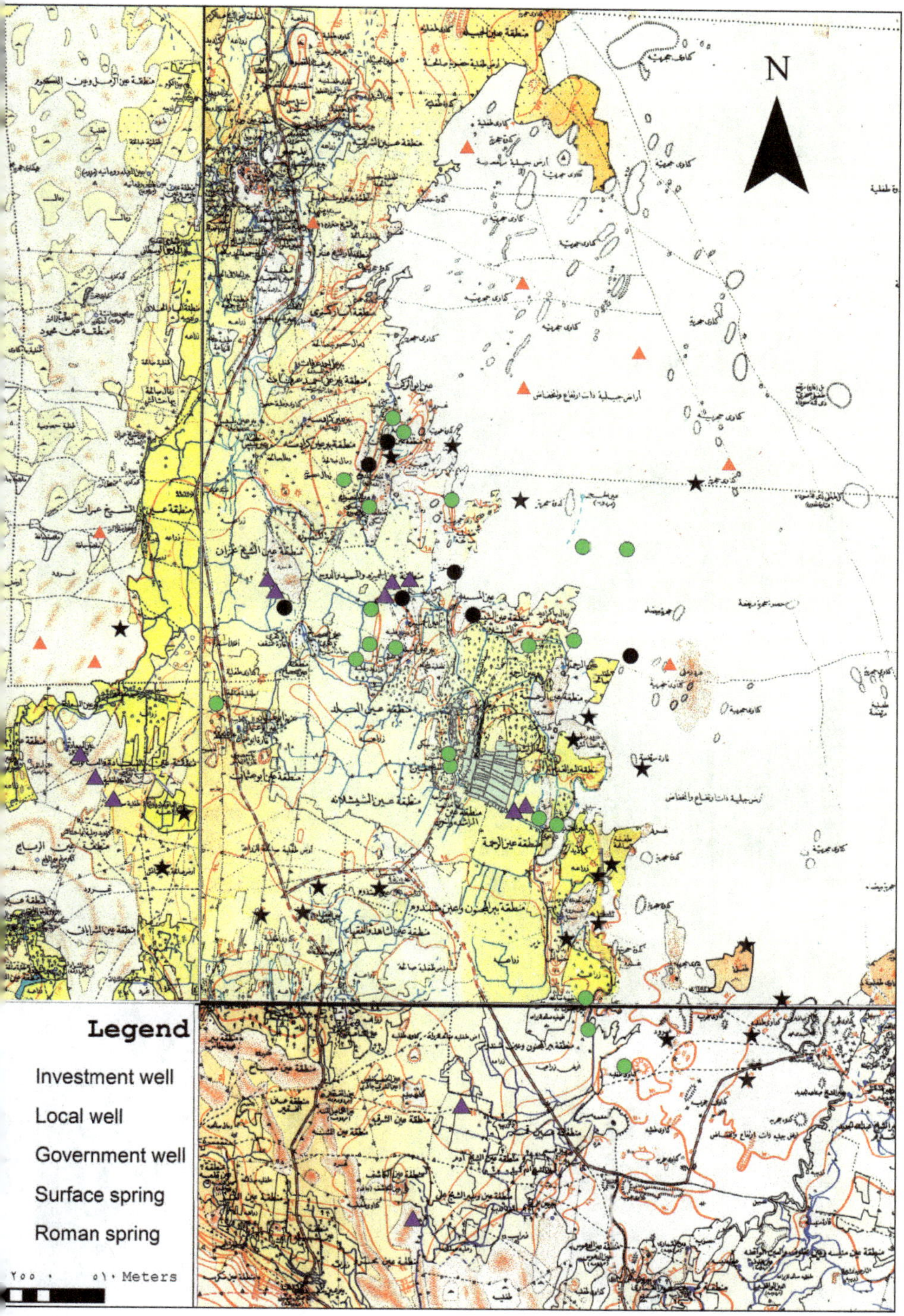

MAP 22 *Location of wells and springs in Rashda (2007). Source*: Kato et al. 2010.

PHOTO 27 *Government well* (bir hukumi)

PHOTO 28 *V-notch weir* (haddar) *to control the water flow*

PHOTO 29 Marwa (*small*) *canal*

PHOTO 30 *Distributor* (maqsam) *of water*

PHOTO 31 Ghamr (*flood*) *irrigation for growing Egyptian clovers*

PHOTO 32 Bakiya *irrigation for growing garlics*

PHOTO 33 *Surface spring* ('ain sathi) *in Well No. 3 Irrigation District*

PHOTO 34 *Investment well* (bir istithmari)

Excursus 4

Bir Majnun District

Bir Majnun (or Bir Qibli Maarif) is a local well located two kilometres to the south-west of the Rashda residential area. It was drilled in 1878, according to a report on the inspection of water for drinking in Egypt Published in 1931 (Azadiyan 1931: 69). It was a shallow artesian well, 120 metres deep, and its water discharge is estimated to be 60 *qirat* (6,000 cubic metres) per day. It irrigates 320 *feddan* of land in winter. The areas irrigated are the Ain al-Jadida, Manaqil al-Sharqiya wa al-Gharbiya, and Hamraya al-Jadida districts.

It was named "Majnun" (meaning "crazy" in Arabic), because it discharged such a large volume of water. The report states as follows.

> When digging was finished, the water flowed from the well in bubbling force. As a result, the well was named "Bir Majnuna", which means "crazy well". The water discharge (*miqdar al-ma' al-munsaraf*) reached 10 *qirat*, and after leaving the well, the water flows as a small stream for a long distance. Trees grow on the banks of the stream. It is a beautiful sight. (Azadiyan 1931: 69)

The well discharged plenty of water, so the owner of the largest share of the water, the umda, gave landless farmers the right to access an amount of water. However, the water ceased to flow around 1997. An electric pump was installed, but it discharged only a small amount of water. Therefore, Well No. 11 (or Bir Qibli Mukhaiem, meaning well west of the camp) was drilled in 2004 in the Duhus area near the Bedouin Camp Hotel (see Map 23). Initially it was an artesian well and discharged about 60 *qirat* of water per day. However, it stopped discharging water after three years, so it was replaced by a deep well, 1,050 metres deep, equipped with a diesel-powered pump.

In 1997, the owners of water rights decided to make a new canal because the sand flowed into the existing canal. The chief of the Local Unit of Rashda at the time, who was also one of the shareholders, went to Kharga and met the governor and the director of the Rural Development Authority to obtain material and financial support such as cement, funds to transporting stones, and similar aid.

After obtaining an accord for financial and material support, the four chief water sharers formed a water management board (*majlis al-nizara*). Under this board, the new canal was created in the current location. At the same time, the board reclaimed the land by "putting their hand" and allotted the water to 14 people who contributed to the construction of the canal. Now this area is called "Jaala Gharbiya" because it is especially reserved for these 14 people as compensation (jaala) of their contribution.

The division of the water supply for these 14 people was decided in the following way. First, one "share" (*sahm*) was set at six hours of irrigation. Then the share for each person was decided according to their contribution to the work as shown in the Table 38. There are a total of 22.5 shares, and because it was agreed that each share would be six hours, the total is 135 hours of irrigation, corresponding to 16 *amila*, if an *amila* is defined as eight hours.

TABLE 38 *Time share of water supply in Jaala Gharbiya district*

ID	Share of jaala	Type of contribution	Share of board	Total	Hour	*Amila* (8 hours)	*Amila* (12 hours)	Share of *amila*
1	2.5	Provided workers	0.5	3.0	18.0	2.3	1.5	1.1
2	3.0	Provided workers		3.0	18.0	2.3	1.5	1.1
3	2.0	Provided workers		2.0	12.0	1.5	1.0	0.7
4	1.0	Provided workers	0.5	1.5	9.0	1.1	0.8	0.5
5	1.0	Worked himself		1.0	6.0	0.8	0.5	0.4
6	2.5	Worked himself		2.5	15.0	1.9	1.3	0.9
7	1.0	Worked himself		1.0	6.0	0.8	0.5	0.4
8	1.0	Worked himself		1.0	6.0	0.8	0.5	0.4
9	0.5	Worked himself		0.5	3.0	0.4	0.3	0.2
10	2.0	Worked himself		2.0	12.0	1.5	1.0	0.7
11	0.5	Worked himself		0.5	3.0	0.4	0.3	0.2
12	1.0	Worked himself	0.5	1.5	9.0	1.1	0.8	0.5
13	1.5	Worked himself		1.5	9.0	1.1	0.8	0.5
14	1.5	Worked himself		1.5	9.0	1.1	0.8	0.5
Total	21.0		1.5	22.5	135.0	16.9	11.3	8.0

Source: Information collected by informant in 2012.

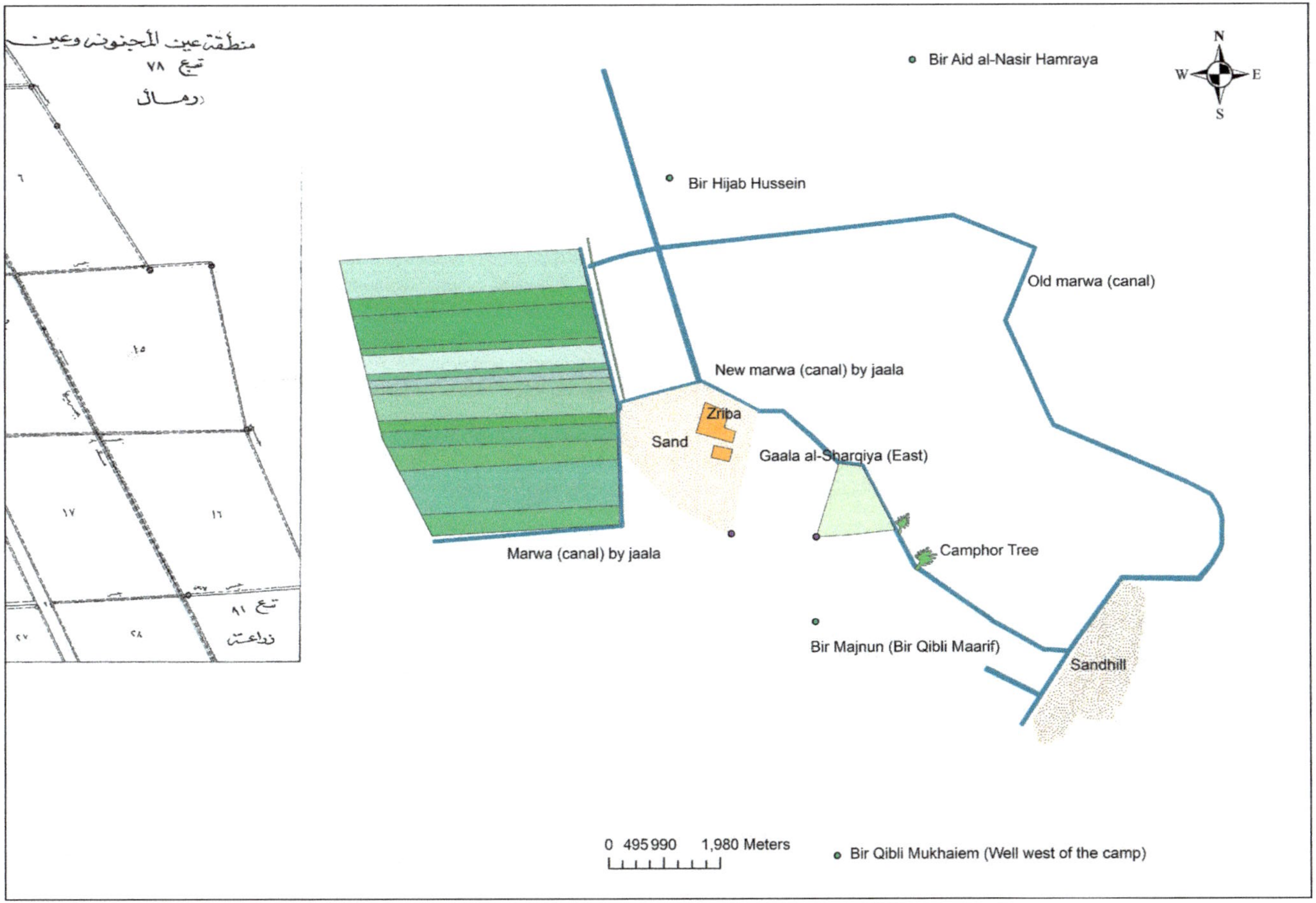

MAP 23 *Location of Bir Majnun and the Jaala Gharbiya district. Source*: Information collected by informant in 2012 and fieldwork in 2013.

CHAPTER 11

Agriculture and Household Economies

Despite the peri-urban character pointed out in Chapter 4, the economic life of oasis villages, in this case Rashda, is basically dependent on agriculture. This chapter describes the economic activities related to agriculture in the household and market economies in Rashda today, the relationship between well and land distribution, the combination of non-farm, farm and livestock activities, the kinds of crops and livestock, crop rotation, and the marketing of crops, using the household survey data collected in 2005 and 2009.[1]

Agriculture in Household Economies

Agricultural Practices in Rural Egypt

Although agriculture is less commonly practised in Egyptian villages, it still plays an important role in the rural economy. This can be understood by looking at farm households according to various definitions, using the household survey data obtained from 19 villages in 2005, presented in Chapters 3 and 4.

In the agriculture section of the questionnaire, the householders are asked whether they cultivate land. To this question, an average of 48.8% of the householders belonging to 19 villages said "yes". Landless people (without cultivated land) comprise an average of 51.2% of households in each village. This proportion differs only slightly between villages in the Nile Valley and the oasis region.[2]

An interesting fact is that there is a discrepancy in the importance of agriculture according to the criteria of farm households. The proportion of farm households rises to an average of 72.7% when they are classified according to the criterion of receiving income from farm activities.[3] In most villages, households practise farm and non-farm activities together. Because the income from

1 Detailed information on landholding in Rashda was collected by the questions in the household survey conducted in 2009 on the plots cultivated by the households.

2 Except for Liwa Subih (86%), which is a newly created "village of graduates" promoted by government to create employment in agriculture.

3 Borr Bahry and Liwa Subih have an especially high proportion of households living solely on farm income. Zahra, Sidi Oqba, Homa, Beni Samrag, Kawamel Bahry, and Awlad Sheykh villages also have more than 40% of their households living solely on farm income. For other villages, the proportion of households living solely on non-farm income is less than 30%.

 | DOI 10.1163/9789004317390_013

farm activities is less than that from non-farm activities, most of these "mixed" households seem to practise farming as a secondary activity.[4]

Another approach to farm households is to classify their members aged 15 years and over by the number of days that they have worked in agriculture during the 12 months prior to the study. An average of 48.3% of male members and 81.5% of female members reported having worked in agriculture for more than 200 days. The percentage of female members is considerable, because female members tend either not to work or to work in agriculture. At the household level, 66% of households have at least one member working in agriculture.[5]

Why is there a discrepancy between the definitions of farming households? One reason lies in the practice of animal husbandry. In Egypt, where all land is irrigated either by the Nile or by underground water, the mixed agriculture–livestock system represents the majority of farming in the Nile Valley and the oasis region. Many farmers raise large ruminants (buffaloes and cattle), small ruminants, and poultry while growing Egyptian clover (*birsim baladi*).

However, it should be noted that the importance of livestock and the combination of the agriculture–livestock system differs by region.[6] The income earned from livestock is especially important in the villages of Dakhla Oasis. The income derived from livestock as meat is about three times more there than in the villages of the Nile Valley. Around 40% to 50% of the households in Rashda and Balat villages and the majority of those in Liwa Subih own cattle.

Another reason for the discrepancy could be the existence of "family land". There is no specific word for "family land" in Egypt, but we used this word and asked the households about ownership or sharing of such land, assuming that there were many villagers who would inherit it and cultivate their parents' land.

The proportion of households that have "family land" is an average of 40%. This proportion is higher in the Nile Valley than in the oasis region.[7] This may be because of fragmentation in the old land. However, it will be shown in detail below that shared cultivation of "family land" is frequent in Rashda village, and this represents how families function in household economies.

4 Liwa Subih and Borr Bahry are special cases where more than 60% of the households are earning income solely from fishing and cultivation.

5 This proportion varies by village, but the four villages of Bulaq, Rashda in the oasis region, Abu Senita in central Lower Egypt, and Sheykh Issa in southern Upper Egypt tend to be low.

6 As to the villages in peripheral Lower Egypt (Hamra, Kafr Beni Hilal, and Sidi Oqba), Kafr Shubrahur in Central Lower Egypt, and Middle Egypt (Damashqin, Beni Samrag), they grow cattle and buffaloes mainly for producing milk and butter.

7 It is especially high in Zahra and Abu Tawala villages in Sharqiya governorate, and Kawamel Bahry and Awlad Sheykh in Suhag governorate.

Importance of Agriculture in Income Stratification

The importance of agriculture in Rashda may be explained, first of all, by the income class structure. Table 39 shows the structure of estimated household income per capita.[8] The main income source in Rashda is non-farm wage employment. This accounts for half the income. Income derived from self-employed farming, on the other hand, accounts for only 29.7% of total income.

The average household income per capita was 7,515 Egyptian Pounds in 2009, but there is a wide disparity between the lowest and highest income categories. It was found that those in the highest income category earn six times more than those in the lowest income category.

This wide disparity is caused by the distribution of farm income. When income sources are classified according to class, it is noticeable that self-employed farming is the most popular source of income for the richest income groups. Thus, although farming contributes little as a source of income, it has a substantial impact on income stratification in Rashda.

Land Distribution

Obviously, the determinant factor of the income distribution is land assets. Land distribution is very uneven in rural Egypt, and this also holds true for Rashda and the oasis region in general. Indeed, there is a large disparity in landholdings between the villagers in Rashda. Of the surveyed households, 58.5% are landless. The mean area of cultivated land in Rashda is 1.1 *feddan*.[9] Half the landholders have less than two *feddan* of land, while 18.2% of the landholders have five *feddan* or more. Some of the few large landholders have extremely large landholdings: the largest landholder has 450 *feddan*, followed by another with 150, and another two with 42 *feddan*.

As in the Nile Valley, household income clearly increases with size of landholding. Moreover, landless households belong to the low-income category. The proportion of the highest income category among the households holding six or more *feddan* is 66.7%, compared with only 12.4% for landless households.

8 See the definitions of incomes in the footnote 4 (page 5) in Chapter 4.

9 It should be noted that the area of cultivated land in Rashda and the oasis region in general is larger than that of other surveyed villages in the Nile Valley. The mean area of cultivated land in the villages in the Nile Valley was half a *feddan*, whereas it was one *feddan* or more in the Oasis villages. This may be because access to water is a determinant factor, and land is available.

TABLE 39 *Household income level and components (percentage share of the overall per capita income)*

Income categories	Per capita income (LE)	Self-employment				Wage		Other			Total
		Farm	(Land)	(Livestock)	Nonfarm	Farm	Nonfarm	Real estate	Financial	Transfer	
Lowest	3,135	17.8	(9)	(9)	2.8	2.0	35.8	16.6	0.4	24.5	100.0
Second	4,031	17.8	(9)	(8)	8.9	2.2	49.6	10.9	0.3	10.4	100.0
Third	4,927	17.1	(10)	(7)	3.6	2.5	58.9	8.9	0.1	8.9	100.0
Fourth	8,020	27.2	(17)	(10)	7.2	1.2	45.2	9.7	0.0	9.5	100.0
Highest	17,461	41.8	(31)	(11)	10.4	2.6	31.9	7.1	0.6	5.6	100.0
Total	7,515	24.3	(15)	(9)	6.6	2.1	44.3	10.6	0.3	11.8	100.0

Note: There are 715 households in the sample.
Source: Household Survey 2009.

Then who has land and who does not? A village that has experienced an inflow of agricultural workers is likely to give the impression that old families such as the umda's own the land, while outsiders remain landless workers. Indeed, seven households that hold six or more *feddan* belong to the umda's family of Abd al-Hafiz. However, in reality, there is no statistically significant difference between the landholdings of families. The differences are because there is a wide disparity within families. In fact, half of the households belonging to the family of Abd al-Hafiz are landless. Another reason is that some households that do not belong to old families are large landholders. For instance, the largest landholder, with 450 *feddan*, and another with 150 *feddan* are both originally from outside the village. Their families came to work under the umda in the past.

Farmers gain access to cultivated land in four main ways. The first way is by inheritance from their fathers (or sometimes mothers) under Islamic inheritance rules.[10] In Rashda, 49% of the farmers obtained the land through this type of access. The second way is the purchase of land from the government or another villager—19.7% of the farmers obtained their land in this way. The third way is renting or share-cropping, practised by 20.1% of the farmers. As for large landholders, the majority obtained the cultivated land by purchase, or by renting it from the government or from villagers, or by share-cropping.[11] They bought or rented the land from the government as newly reclaimed land, or they purchased it from a villager.

The New Valley Project

In Rashda and in oasis villages in general, access to land is closely related to access to wells, because land has value when there is water drawn from underground. In this regard, the New Valley project implemented during the late 1950s and 1960s had a substantial impact on land distribution. As explained in Chapter 1, the government wells were drilled with a double aim. One aim was to increase the production of staple crops—namely, wheat—for food security, and another aim was to assist land reform under a socialist regime that aimed

10 According to the household survey conducted in 2009, 71.4% of cultivated land is owned by the cultivators, 18.0% is rented, and 10.6% is share-cropped. Half of the latter is share-cropped in return for half of the produce.

11 For the access to the cultivated land, we asked whether the owner of the plot was either a household member, from the same family/a relative, from the same village, or other, and asked what the relationship with the household head was if the answer was "from the same family/a relative". For the share-cropped land, the most common situation is share-cropping of the land owned by the farmer's father, followed by ownership by a villager. The most common owner of rented land is the government.

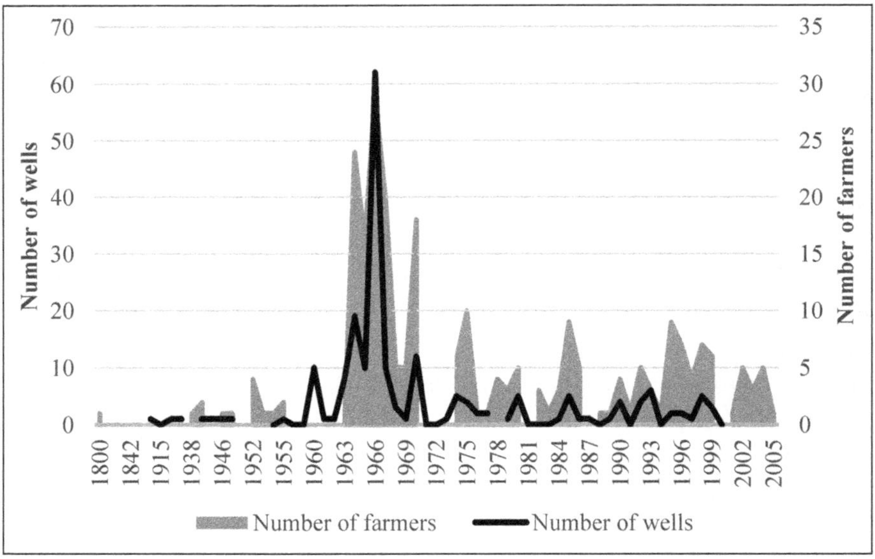

GRAPH 3 *Year of establishment of wells (number of households) (2005). Source*: Household Survey 2005.

to improve the livelihood of Egyptian peasants after the July Revolution in 1952.

As described in detail in Chapter 13, land attached to government wells is leased to the villagers on the condition that they are landless and in low-income households. According to an informant, this had a considerable effect on upgrading the living standards of households in Rashda.

The importance of this project can be understood from Graph 3. This shows the year of establishment of the wells to which farm households are attached and the year when they started cultivation. The majority of the wells were constructed in the 1960s, and this is when the households began cultivation.

Combination of Non-farm, Farm and Livestock Activities

Most households in Rashda make their living from a combination of farm and non-farm activities. In fact, only 17.6 of households lived solely on non-farm incomes, and 10.5% lived solely on farm income in 2009. The majority of households either farmed as a secondary activity (52.7%) or as their primary activity (19.2%). Even among the landless households, only 30.1% lived solely on non-farm incomes.

As pointed out in the previous section, animal husbandry is one reason for the discrepancy between farm households according to the criteria of cultivation and in farm income. Raising livestock is an important agricultural activity and produces 9% of household income. It is important for any income category, unlike income derived from land cultivation, which has become much more important for those in higher income categories.

It is noticeable that those in low income categories secure their livelihood by diversifying the economic activities. Livestock is an important asset for this household strategy, because it provides income, unlike land, which those on low incomes lack.

The type of livestock is similar for all income categories. As mentioned in the previous section, this is cattle. They are mostly raised for milk, which is generally for household consumption, and meat, which is usually sold to traders.

Importance of Family Agriculture

Of the households in Rashda, 24.7% reported having "family land". Ownership of "family land" is related to age. In fact, 35% of household heads younger than 40 have "family land" compared with 15% among those aged 55 or older.

Landholdings are related to age in the opposite direction. Older household heads are more likely to hold land than younger ones. This means that agriculture follows family cycles. Agriculture is managed by the older generation, while the younger generation undertakes non-farming activities. Although the economic unit is the household, the households of fathers, sons, or brothers may co-operate to optimize their labour distribution, under Islamic inheritance laws.

This is a family strategy to optimize utility by deploying different economic activities to maintain living standards. In this regard, the relative proportions of working days in farming and non-farming activities of each household member provide interesting information on how households conduct various activities.

Table 40 shows the employment structure based on the days worked during the previous 12 months for those who reported working at the time of the study. Only 40.2% of respondents had worked only in farming. The percentage is even lower for male workers. Half of the male workers work only in non-farm activities. In contrast, the female workers are much more likely to work on farm activities, although the women tend to be less active than the men.

In terms of age, both men and women of the older generation tend to engage in farming, whereas the younger generation works in non-farm activities. Thus, women, older men, and fathers are in charge of daily farm activities. When more labour is required, such as seeding or harvesting, the households call upon reciprocal labour, *musharaka*, as described in Chapter 8.

TABLE 40 *Employment structure based on the number of days worked over 12 months (%) (2009)*

	Female	Male	Total
Only farming	62.2	25.2	40.2
Only waged agricultural labour	0.0	2.3	1.4
Only non-farm work	24.6	49.8	39.6
Both farming and waged agricultural labour	0.2	0.6	0.4
Both farming and non-farm work	11.1	19.9	16.3
Both non-farm and waged agricultural labour	0.0	0.5	0.3
Mainly farming	2.0	1.8	1.9
Total	100.0	100.0	100.0
(Number)	606	896	1,502

Note: Household members who reported that they had worked during the 12 months prior to the study fieldwork and had engaged in farming activities for 200 days or more per year are classified as "mainly farming", and those engaged for 200 days or more per year in non-farming activities are classified as "mainly non-farm work". Others are classified as "both" ("both farming and waged agricultural labour, both farming and non-farm work, both non-farm work and waged agricultural labour").
Source: Household Survey 2009.

Non-farm Employment Situation in 2009

A household strategy of family agriculture is related to the expansion of non-farm activities in the village. As described in Chapter 4, non-farm employment is the main source of employment in today's rural Egypt. In Rashda, it is difficult to find a genuine farm household making its livelihood based only on farm activities. Most households have one or more members engaged in non-farm activity.

The predominant feature of the employment structure in Rashda is that it is overwhelmingly dependent on the government sector. Among waged workers, 72.1% work in the government sector. Those in the government sector work either in administration or in education. Of the waged workers, 60.3% work inside the village. Others mainly work in the town of Mut (68.3% of waged workers work outside the village) or in the villages in Dakhla. They mostly work in the Local Unit or other government offices, or as teachers in Rashda or nearby villages or in the town of Mut.

The government sector is especially important for women. In Rashda, women work quasi-exclusively in the government sector, which employs 91.7% of female waged workers, compared with 65.2% of male waged workers.

Only 10.6% of households have members who are self-employed in a non-agricultural field. Of the 55 self-employed people, 54 currently work inside the village. Of these, 26 are self-employed in commerce and in shops. Other self-employed people work in transportation, such as driving taxis or trucks.

There is practically no labour migration away from the village: 15.2% of the households reported that one member had lived or currently lives outside Rashda for work. Their destination was Cairo (37.4%), Giza (13.9%), or Matruf (20.0%). Eighteen of 115 people have worked or currently live abroad: 10 in Saudi Arabia, four in Libya, and four in other Arab countries (Jordan, Kuwait, Qatar, and Oman). This migration abroad occurred during the 1980s and 1990s, except for three people currently living in Saudi Arabia.

Of the waged workers who were working at the time of the study fieldwork, 65.4% found their jobs through the government (Employment Office). Very few found their current jobs through their families (13.9%) or by contacting employers directly (19.9%). Thus, we may point out that non-farm job opportunities are largely created by the government, and the link with the government is very important.

Crops and Livestock

The agricultural season is divided into two seasons. The first is the winter season from November to April (or May). The second is the summer season, from May to October. Like other villages in Dakhla Oasis, the most common crops in Rashda are wheat and Egyptian clover (*birsim baladi*) in winter, and rice and sorghum in summer, while dates and alfalfa (*birsim hijazi*) are perennial.

Rice has been the preferred summer crop for many years, even though it requires more water than any other crop in general and particularly in the oasis region with its high evaporation rate.[12] It seems strange to see rice fields in an oasis, where the water is not abundant. However, according to Beadnell (1909), rice has been grown for a long time in the oases, and for good reason; the rice grows on land containing a proportion of salt and provides a periodical cleansing of the land, which would otherwise become so salty that it would

12 There are good reasons for the choice of rice as a crop: it is the most profitable staple crop, and it is well adapted to adverse summer conditions. Second, farmers are aware that growing rice is the best means to leach down the continuously accumulating salts in the soil. Third, leaving the land fallow during summer would facilitate salt accumulation, and as a result, it is good for harm-sensitive winter crops such as wheat and Egyptian clover (Montaz and Siddiq 1989: 280).

adversely affect other crops grown on the same ground (Beadnell 1909: 214). However, in the New Valley governorate, rice production is forbidden by law to reduce overdraft of water from aquifers.[13]

Dates are one of the most important agricultural products. Date palms are suitable in dry and rainless areas, but they do not yield fruit unless their roots are well watered. For this reason, dates are best suited to oasis agriculture. There are 57,417 date palms in Rashda (2007). There are many varieties of date palms, but in Rashda, and in the Western Desert in general, the most common varieties are *Saʿidi* and *Tamar*.[14] *Saʿidi* dates are a popular variety in Egypt for their soft and sweet taste, and are sold in markets mostly through the date factory operated by the Agricultural Co-operative in the town of Mut, where dates are cleaned, packed, and exported to Cairo, or to European countries. *Saʿidi* dates are also used to make a popular compressed paste of dates, *ʿagwa*, which is used for making sweets such as *maamul*. *Tamar* is a variety usually eaten after sun drying, and mostly for home consumption.

Egyptian clover is a fodder crop for cattle. It is a common crop in the Nile Valley but also in the Western Desert. It has a deep green colour and has white flowers in winter from November to May. It is used for a double purpose, as livestock fodder and as fertilizer. It is rich in nitrogen, and it enhances crop production from the soil. As such, it is called the "king" of fodder crops, and many farmers use it as a half-yearly rotation crop. Egyptian clover grows quickly, so it can be cut four or five times in a season. It is used particularly after rice cultivation, to render the saline and poorly drained soil fertile.

There are five varieties of Egyptian clover in Egypt: *misqawi*, *saʿidi*, *fahl*, and *khadrawi*. In Rashda, the most common variety is the *misqawi* type. It is common in Lower Egypt, where there is abundant water in winter; it grows rapidly and can be cut four to eight times in a season (Kennedie and Mackie 1925: 13).

Alfalfa or lucerne is a common crop to provide fodder for cattle. If Egyptian clover is the "king" of fodder crops, alfalfa is the "queen". It has less nitrogen than Egyptian clover, but it is much more suitable for dried hay. Probably for this reason, alfalfa is exported to Gulf countries in bales. Every day, many trucks take alfalfa bales from Dakhla Oasis for export. However, probably because the farms in Rashda are of a small scale, most of the alfalfa is sold in local markets. In Rashda, alfalfa is planted for three consecutive years.

---

13 See "Irrigation Ministry Prevents Rice Cultivation in Deserts" in the *Egypt Independent* 2014/4/24 <http://www.egyptindependent.com/news>.

14 *Hayani* is also a common variety in Egypt, but it is rarely grown in Rashda. It is red-black in colour and is mostly for domestic sales rather than export.

The area where each of these main crops is cultivated is shown in Table 41. Vegetables such as onions, eggplants, okra, green peppers, and fruits such as melons and watermelons are grown, but there are no data on them. This is because the data collected by the Agricultural Co-operative only cover cash crops, and the vegetables are generally used for household consumption.

The crops cultivated differ by type of well. Alfalfa is a preferred crop in the New Valley governorate in general but more so in districts irrigated by investment wells and surface springs. Dates are generally produced on land attached to local wells and investment wells. The land attached to government wells, on the other hand, does not have palm trees. On such land, the main crops are rice, sorghum in summer, and wheat and Egyptian clover in winter, and alfalfa is a perennial crop.

These crops are also cultivated on land attached to investment wells. In addition, vegetables such as fava beans (*ful*), onions, cow-peas (*lubiya*), and maize are grown there.

TABLE 41 *Area (feddan) of land for cultivation of main crops (2008)*

	Crop	Government and local wells	Investment wells	Surface springs	Total
Winter	Wheat	750	332	84	1166
	Alfalfa (*birsim hijazi*)	360	285	200	845
	Egyptian clover (*birsim baladi*)	400	89		489
	Barley	12	18	4	34
	Fava beans (*ful*)	6	20	3	29
	Egyptian lupin (*tirmis*)	3		2	5
	Peas (*bsila*)	1		2	3
	Onion	4	14	12	30
Summer	Rice	150			150
	Cow-pea (*lubiya*)	35	35		70
	Sorghum	50	10		60
	Maize	10	15		25

Note: One *feddan* equals 0.42 ha.
Source: Agricultural Co-operative in Rashda, information from an informant in November 2008.

Crop Rotation in the Government Well Irrigation Districts

The major reason for the difference in crops between the land attached to government wells and other wells is that crops from government wells are managed by an agronomist attached to the Agricultural Co-operative. Each year a meeting with agronomists is held to decide which crops to cultivate. Under the supervision of the Agricultural Co-operative, the farmers of the government well cultivate the same crops under a crop rotation system (Table 42).

This is a system that has generally been used in the Nile basin, especially when rice cultivation was controlled by the government until the 1990s. According to this system, farmers hold three plots each in different subdistricts. Because rice cultivation requires permanent irrigation, and water flows for other crops must be controlled to avoid the overuse of water, crop rotation is necessaryKinds of crops and their rotations are decided by the Directorate of the Ministry of Agriculture. One of the three subdistricts must cultivate alfalfa for three years, while two subdistricts participate in crop rotation as in Table 42. For this reason, crops are rotated between subdistricts every three years and six months.

No meeting is held to decide when and what kind of crops to cultivate, because the same cropping pattern is followed every year. However, in some circumstances that require the agreement of cultivators, such as the prohibition of rice cultivation, the Agricultural Co-operative assumes the task of informing and directing the cultivators.

As Hansen explains, crop rotation in the Nile basin is intended to combine the advantages of private ownership and small farmers' initiatives with economies of scale in irrigation, soil preparation, financing, and trading, and the system gives individual farmers land in different parts of the area under common rotation (Hansen 1975: 151).[15] The three-year rotation system would thus imply that farmers should be given land in three different places, each with a different crop.

In this way, large areas can be grown with the same crop, while the individual small farmer on his three plots of land will apply the same rotation that he would apply individually and will always have a food crop for his family, a feed crop for buffalos, and a cash crop for his expenses. Soil preparation, including ploughing and irrigation, may become more rational, taking advantage of economies of scale in these processes. Mechanization becomes possible, whereas seeding, weeding, fertilizing, and harvesting are left to the individual

15 The system was originally invented by the British during the inter-war period and applied to the Gezira project in the bifurcation of the White and Blue Nile in the Sudan, whence the Egyptians adopted the idea (Hansen 1975: 151).

TABLE 42 *Example of crop rotation in government well districts*

	Subdistrict A	Subdistrict B	Subdistrict C
Winter	Wheat	Egyptian clover	
Summer	Rice (or sorghum)	Fallow land	
Winter	Egyptian clover	Wheat	
Summer	Fallow land	Rice (or soghum)	Alfalfa for three years
Winter	Wheat	Egyptian clover	
Summer	Rice (or sorghum)	Fallow land	

Source: Information from informant and Agricultural Co-operative.

owner. Seeds and fertilizers are provided by the Agricultural Co-operative (Hansen 1975: 150).

Livestock

Rearing livestock is also an important agricultural activity in Rashda. From our observations, it became apparent that a large percentage of crops were Egyptian clover and alfalfa, which are animal fodder. A sizeable portion of Rashda is dedicated to housing animals—cows, goats, sheep, and horses comprise the largest animal populations in the constructed animal sheds (*zariba*) that are seen in many places surrounding the residential area.

However, unlike areas in the Nile delta and valley, water buffalos—a primary animal in the Nile delta for producing milk and as draft animals before mechanization—are rarely raised, because they require more water than cattle, and a wet climate. In contrast, cattle are raised for meat and milk production. Milk is used for drinking. Cheese is rarely produced in Rashda because of its hot climate.

Market in Household Economies

Crop Production in 2009

The crops are consumed by households, sold, or given to the landowner. Table 43 shows the structure of the usage of main crop production. Only a small number of farm households reported that they had produced vegetables or fruits other than dates (such as mangos).

Half of the products are consumed by the households and contribute to maintaining the household consumption level. More wheat is consumed by households than rice or dates. Fodder crops—Egyptian clover and alfalfa—are also used for household consumption. Moreover, 76.1% of the fodder crops produced by cultivators are used to feed their animals.

Crops are also exchanged at market or between villagers. Small cultivators with less than one *feddan* grow wheat and Egyptian clover, essentially for household consumption. In contrast, medium cultivators (with two *feddan*) and large cultivators tend to cultivate different crops for market or for the landowner. For instance, the proportion of household consumption of wheat among cultivators of less than one *feddan* was 78.5%, compared with 36.9% among cultivators of two *feddan*. For the latter, the remainder of the wheat crop is sold at market (45.6%) or given to the landowner (17.5%). Landowners either can be tenants of the land at a fixed price or can share the produce. In Rashda, the latter arrangement is more common.

The crops can be sold either to merchants, to an Agricultural Co-operative, or directly at local markets or to other farmers. The characteristic feature of agricultural marketing in Rashda is that the Agricultural Co-operative plays an important role in marketing. In the case of wheat, 64% of cultivators sold to the co-operative. For dates also, 59.0% of the cultivators sold to the co-operative or to the date factory managed by the co-operative in Mut city. For rice, 47.9% of the cultivators sold their produce to the Agricultural Co-operative,

TABLE 43 *Consumption of the main crops produced in Rashda (%) (2009)*

	Consumed by household	Given to landowner	Sold	Total	Number of households
Rice	35.3	15.5	49.2	100.0	128
Wheat	52.7	14.1	33.1	100.0	279
Sorghum/Maize	53.6	20.8	25.6	100.0	10
Vegetable	52.2	7.4	40.4	100.0	9
Date	38.1	12.8	49.1	100.0	83
Fruit	6.9	51.9	41.3	100.0	3
Fodder crop	76.1	14.6	9.3	100.0	225
Straw	77.3	13.2	9.6	100.0	262
Other crop	51.1	14.9	33.9	100.0	23

Notes: 1. Proportions were calculated based on the quantity produced; 2. Vegetables include eggplants, cucumbers, molkheya, okra, and beans (broad beans, lentils, etc.).
Source: Household Survey data 2009.

and the other half sold to merchants. Although marketing is free for all crops, including rice and wheat, which were under governmental control until the 1980s, the Agricultural Co-operative still plays a role in marketing in Rashda.

There were no sales of produce at the local market. This is probably because there is no weekly market in the village. Merchants are usually from outside the New Valley governorate, such as from Assyut city in Upper Egypt. Merchants from Cairo also come to buy dates and alfalfa.

PHOTO 35 *Tamar date palm trees,* sa'idi *date palm trees (bottom right), and guava trees (bottom left)*

PHOTO 36 *Harvesting dates* (sa'idi)

PHOTO 38 *Packing of date products in factory in Mut*

PHOTO 37 *Date factory in Mut*

PART 4

Community in Well No. 3 Irrigation District

CHAPTER 12

Irrigation Districts

The agricultural land in Rashda is divided into fields. They are called "irrigation districts", since they are assigned to a specific well or spring. The system of irrigation and the customs of cultivation in the irrigation districts differ according to the type of well or spring from which the water comes. This chapter defines the categories of irrigation districts, and then, focuses on four irrigation districts, two irrigated by a local well, one by an investment well and one by a surface spring, and examines their history, water and land ownership, water distribution and management, water trading between farmers, cultivation system, and so on.

Categories of Irrigation Districts

In Rashda, agricultural fields are assigned to a numbered well. We call such agricultural fields "irrigation districts". They represent the spaces where farmers cultivate land, implement crop decisions, interact with neighbouring farmers, graze their livestock, and ultimately utilize their allocation of the water supply.

In addition to denoting a particular agricultural area, wells are also points where farmers assemble and water is allocated. At each well, there is a mechanism that serves the collective of growers through the distribution of an irrigation schedule.

As mentioned in Chapter 10, the wells in Rashda can be categorized into four types: local wells (*bir ahli*), investment wells (*bir istithmari*), surface springs (*ʿain sathi*) and government wells (*bir hukumi*). The management of the irrigation districts differs according to the type of well.

This chapter focuses on typical forms of management of the three types of irrigation district: those irrigated by a local well, an investment well, or a surface spring. The districts irrigated by government wells will be discussed separately in Chapter 12, because such districts differ from others through the intervention of the state in water affairs.

 | DOI 10.1163/9789004317390_014

Bir Sheykh (Local Well) Irrigation District

Bir Sheykh Irrigation District is an area irrigated by local wells. It is located west of the old residential area (Map 24) and is composed of a narrow strip of land called Bir Nakhil and a flat field called Bir Sheykh.

Bir Nakhil is an orchard of date palms, as the name "Bir Nakhil" suggests. The area is one of the districts of palm trees in Rashda irrigated by local wells (see Table 33 on page 183). The area is about 52 *feddan*, which had 8,878 palm trees in 2008. Except for some animal sheds (*zariba*) for goats and sheep, storage buildings (*shuna*), one mosque (Izba al-Qara Mosque), and one mill (*tahuna*), the area is entirely covered with palm trees. Like other irrigation districts in Rashda and Dakhla Oasis in general, the palm trees planted in Bir Nakhil are of the *saʿidi* and *tamr* types (Table 44). Some fruit trees, such as mango and orange trees, have also been planted.

Bir Sheykh is a flat field located south of Bir Nakhil. Bir Sheykh has an area of 302.5 *feddan*, and it is divided into 42 plots owned by 42 farmers. Although this District has a proper boundary, it is considered to be an annex to Bir Nakhil because the water allocated to Bir Sheykh comes from the well in Bir Nakhil, "Bir Sheykh Husayn". Farmers cultivate rice or maize/sorghum in summer (from May to November) and wheat or Egyptian clover (*birsim baladi*) in winter.

There is no clear explanation for the origin of the name, but one explanation is that the name "Shaykh Husayn" comes from the name of a saint whose sanctuary is located at the edge of the village.

History of the Wells

The land in Rashda has been irrigated for a long time, probably since before the 1800s by water drawn from the Roman springs, Ain al-Shishlana and Ain al-Musid.[1] Water from these two springs continued to flow until 1980; an informant remembers that these two springs had water until that date. However, the quantity of water flow decreased to a critical level, so Bir Sheykh Husayn was dug in 1914 (see Map 24). This is the oldest well in the District and is currently called Sheykh Husayn al-Qadim (meaning "old").

Another well was dug around 1950, because the water discharge from Sheykh Husayn al-Qadim became very low. Because it replaced a dried up well or one with little water, it was called Bir Sheykh Husayn al-Badil (meaning "replacement").

1 See Chapter 5.

TABLE 44 *Number of trees in the date palm orchard (Bir Nakhil) in Bir Sheykh district (2011)*

		Number of trees	Average per plot
Palm trees—fruitful	*Sa'idi*	941	4.8
	Tamar	513	2.6
	Other dates	246	1.3
Palm trees—unfruitful	*Sa'idi*	378	1.9
	Tamar	97	0.5
Other fruits	Mango	131	0.7
	Orange	35	0.2
	Other	14	0.0

Note: "Other" fruits are olive trees and fig trees. Palm trees that are 'unfruitful' are mainly young trees that cannot yet yield fruits (dates).
Source: Information collected by informant in 2010–2011.

TABLE 45 *Wells in the Bir Sheykh Husayn district (2015)*

Name of well	Year dug	Depth (m)	Water discharge (m^3/day)	Current state
Sheykh Husayn al-Qadim	1914	120–150		Dried up in 1999
Sheykh Husayn Dirdir	2003 (Substitution of Sheykh Husayn al-Qadim)	120–150		In operation (installed a pump in 2013)
Sheykh Husayn al-Badil	around 1950–1953	120–150		Dried up in 1999
Sheykh Husayn al-Hali	2009	1000	600 m^3/day (operates 24 hours 360 days)	In operation

Source: Information collected through fieldwork in March 2015.

Both of these wells dried up around 1999 and were replaced by another well, Bir Sheykh Husayn Dirdir, dug in 2004. It was initially an artesian well, but because of the lowering of the water level, a diesel-powered pump was installed in 2013. Another well, Bir Sheykh Husayn al-Hali (meaning "current"), was dug in 2009. Currently, these two wells remain in operation.

These two wells were dug by the Irrigation Office of the Ministry of Irrigation and Water Resources in Mut under a contract with a drilling company. Since 1960, when local wells have dried up, the Irrigation Office has been responsible for digging and installing pumps. Wells dug as substitutes are called "substitution wells" (*bir al-istiʿawad*). The latest well of Bir Sheykh Husayn al-Hali taps water at a depth of around 1000 metres and has a government recorded rate of discharge of 600 m³/day (Table 45).

Water and Land Ownership

The term "old district" implies that water and land property are complex. Usually, people who own land that requires irrigation own water. A change of water rights should depend on a careful assessment of their existing rights and their contribution to water management including digging wells.

Water quotas are inherited from grandfathers who participated in digging a well. In the case of this district, water use rights were already defined according to the land formerly irrigated by the water from two Roman springs, Ain al-Shishlana and Ain Musid.[2] Therefore, when Bir Sheykh Husayn al-Qadim was dug in 1914, the water quota was decided on the basis of an assessment of existing rights according to the irrigated land area that the farmers already owned and their contributions to digging the new well.

The quota of water drawn from the other three wells follows the initial water rights registered for Bir Sheykh Husayn al-Qadim. Water use rights for each farmer are recorded in a register (*kashf*) kept by the Chief of Water (*raʾis al-maʾ*). If a problem occurs between farmers over water distribution, the Chief of Water resolves it by referring to this register.

A water quota or water right is defined on a time basis. This means that each user has a certain period in which to irrigate his land. In the scheduling process, farmers are assigned a certain number of hours of irrigated water every 15 days. Because this district has an annex of land called Bir Sheykh where rice is grown, and because rice requires much water, 15 days are allocated to the field of Bir Sheykh and two days to the palm tree fields of Bir Nakhil.

Water Management

Besides water allocation monitoring by the Chief of Water, there are many other tasks required to maintain the water allocation, which is essentially a community task because it concerns all the farmers sharing the wells. Then how do the farmers in this district manage collective works?

An interesting finding is that tasks are done on a compensation basis. This basis is a form of co-operative contract called a jaala.[3] The following jaala contracts are known to have existed in the districts irrigated by local wells.

2 Ain al-Shishlana dried up in the 1990s, and Ain Musid in the 1980s (see Chapter 10).

3 On jaala, see Chapter 7.

Nowadays, jaala contracts for canals, transportation, and land preparation are rare.

1. *ja'ala al-nizara* (jaala for water management). The inspector receives *amila* in compensation for inspecting water distribution.
2. *ja'ala al-marwa* (jaala for canal maintenance). The contractor (*muja'ila*) receives *amila* in compensation for the cleaning and maintenance of a canal.
3. *ja'ala al-hirasa* (jaala for a watchman). In the case of dates, a farmer who works as a watchman in an orchard receives a basket of dates from each farmer in compensation for his work.
4. *ja'ala khidma al-nakhil* (jaala for taking care of palm trees). A farmer who takes responsibility for fertilization and other works of palm trees receives one third of production in compensation for his work.
5. *ja'ala saqiya Sidi Husayn Waqf* A farmer in charge of watering the two clay pots served as public fountain everyday receives two palm trees in compensation for his work.

In this District, there is an "(Water) Management Board for the fields of Bir Sheykh Husayn" (*majlis al-nizara li-mazra'a bir Shaykh Husayn*) run by jaala contract. There are four tasks for this board (Table 46). One is the Chief of Water register and allocation, which requires experience in allocating water to the holders of inherited water rights. Usually, the Chief of Water is given this position based upon experience, age, and local prestige. However, in this district, the chief of water is a young man in his 30s. He leases land from the umda and uses the water that he receives as compensation (*amila*) for cultivating rice. The second task is announcing water allocations to the owners at the time of the meeting. The third task is repair of canals, cutting tree roots, and reinforcing the banks (*jisr*). This task is carried out by three farmers every week. The fourth task is providing tractors for making canals, which is done by one farmer every Friday.

These tasks are contracted to farmers by the farmers in the district as jaala, and they are remunerated by water allocation time. The Chief of Water receives one *amila*, and other farmers receive half an *amila*. The exception is the farmer responsible for providing a tractor for the canals. He receives 35 kg of rice as compensation, probably because he has no land or is not interested in cultivation.

Another interesting fact is that water time is used as a form of social welfare. In this district, there are two forms of this. One is that three *amila* are rented to farmers in these fields during the rice season. The produce is distributed to

TABLE 46 *Water administered by Water Management Board* (majlis al-nizara) *of the Bir Sheykh district*

Person ID	Job perfomed	Period of task	Recompensation
1	Water registration and allocation by Chief of Water	Beginning of season	1 *amila*
2	Providing information to the owners	Every time a meeting is held	Half an *amila*
3	Repair of canals, cutting tree roots, reinforcing banks (jisr)	Every week, and every time a problem occurs.	Half an *amila*
4			Half an *amila*
5			Half an *amila*
6	Providing a tractor for canal maintenance	Every Friday	2 *weba* (35 kg of rice)
7	Supporting patients with chronic maladies	Given to farmers during the season of rice cultivation	3 *amila*
8	*Waqfiya* (water reserved for waqf)	Given to the Zakat Committee	1 *amila*

Note: According to the informant, this water waqf exists in more than 90% of the districts irrigated by local wells.
Source: Information collected by the informant in January 2015.

sufferers of chronic maladies. Another form is called Waqf al-Sayyid al-Badawi. Water is considered to be a gift from God, so one *amila* is reserved for welfare use to thank God. The fee for renting one *amila* to a farmer in this district is usually about 500 Egyptian Pounds per season. This amount is collected by the Water Management Board and then donated to the Zakat Committee of the Abu Bakr Mosque Charitable Association. This *waqf* is named after Sayyid Badawi, probably because the Khalwatiya Sufi Order with which the Association is connected is attached to the Saint of Tanta.

Water Trading between Farmers

As mentioned in Chapter 10, water rights are based on the time concept of *amila* and are inherited. Because an *amila* is a unit of relative measurement and not an absolute quantity, and because this district is an old one, there are many farmers who hold the rights to a smaller quantity of water. However, as the history of the wells in the past hundred years shows (see Chapter 10), water flow is uncertain. In such restricted circumstances, how do farmers maximize the efficient use of water and land?

TABLE 47 *Access to water in the rice fields of Bir Sheykh district* (*summer season 2010*)

Plot ID	Area (*feddan*)	Access to water	Water quota			Rice production (kg)
			amila	*qadam*	Total (*qadam*)	
1	15	Inherited	3	15	105	3,900
2	7	Inherited	3	0	90	2,400
3	7	Inherited	1	8	38	1,500
4	7	Inherited	2	0	60	1,500
5	7	Inherited	2	0	60	1,500
6	3	*Nizara* + rented	1	10	40	1,500
7	3					
8	4	Inherited	1	10	40	1,500
9	4	Owned + *nizara*	2	10	70	2,400
10	4					
11	8		4	0	120	4,200
12	6.5	Inherited + *nizara*				
13	5					
14	6	Inherited + *nizara*	1	0	30	1,200
15	11				-	
16	9	Inherited + rented	1	7	37	1,500
17	6	Inherited	1	4	34	1,500
18	5	Inherited	0	6.5	7	300
19	6	Inherited + *waqf*	1	0	30	720
20	4	Inherited	1	0	30	900
21	5	Inherited	1	0	30	900
22	3	Rented	1	0	30	1,020
23	3	Inherited + purchased	1	15	45	1,080
24	3	50% inherited + 50% rented	0	10	10	360
25	8	Inherited	2	0	60	1,980
26	4	Inherited	0	24	24	900
27	5	*Nizara*	0	25	25	1,080
28	2		2	15	75	2,820
29	7	Inherited				
30	7	Inherited	3	0	90	2,400
31	16	Inherited	1	0	30	1,200
32	6	Inherited	2	15	75	2,100
33	7	Inherited	1	15	45	1,380

TABLE 47 *Access to water in the rice fields of Bir Sheykh district (summer season 2010)* (cont.)

Plot ID	Area (*feddan*)	Access to water	Water quota			Rice production (kg)
			amila	*qadam*	Total (*qadam*)	
34	7	Inherited	1	15	45	1,200
35	12	Rented	2	0	60	1,980
36	18	Inherited	6	0	180	4,800
37	5					
38	4	*Nizara* + rented	1	0	30	1,200
39	27	Inherited	3	0	90	3,600
40	3		1	15	45	1,320
41	3	20% inherited + 80% cultivator				
42	28	Rented	1	0	30	900
Total	310.5		53	213	1,810	58,740

Note: Rice cultivation was from 25 May 2010 until 15 November 2010. *Nizara* here means a member of the observatory board.
Source: Information collected by informant in 2010.

One strategy for efficient use of water is water trading. This is especially common in the rice fields where more water is needed than in palm tree fields. As Bedawy (2005: 478) mentioned with reference to Farafra Oasis, users who grow a sensitive crop that needs fewer irrigation intervals may take the water quota of users who grow less sensitive crops. In case of this district, rice is sensitive and needs daily irrigation, but palm trees are less sensitive and can be irrigated once every 15 days. Therefore, rice growers may take water quotas from palm tree growers.

According to information collected in summer 2010 on the Bir Sheykh rice fields (Table 47), there were 42 plots, of which half the farmers had access to water through inheritance of water rights. The rest obtained water either by jaala, or by renting it, or by purchase. Most commonly they obtained additional water by rent or jaala

Renting water can be done on a monetary basis or by exchange. For instance, some farmers exchanged water with storage houses (*shuna*) or animal sheds (*zariba*). We counted 35 storage houses and 10 animal sheds in Bir Sheykh Husayn. Of these, 24 storage houses and seven animal sheds were purchased from, or exchanged with, land owners.

Share-cropping

Share-cropping can be considered to be another strategy for the efficient use of water. Surprisingly, a majority of farmers in this district perform agriculture by share-cropping, mostly by half shares. In the rice field (Bir Sheykh) in Bir Sheykh Husayn district, all the plots except for one are cultivated by share-cropping by half shares between the owner of the land and water, and the cultivator who brings resources such as labour and fertilizer.

In the palm tree orchard (Bir Nakhil) in the Bir Sheykh district, 97 of 178 plots are cultivated by the owners themselves. The remaining plots are cultivated by other farmers. There are some farmers who share-crop three or four plots of different owners together.

According to an informant, one important reason that farmers prefer share-cropping is water scarcity. Because the water allocation is considered to be insufficient, it is not profitable to rent land. For the owner, it is considered to be unreasonable to hire labour and to pay cash for the same reason—that production is insufficient because of water scarcity.

Social Relations in Water Trading

Usually, these two forms of arrangement between farmers—water trading and share-cropping—are founded on social relations based on trust. In this district, these social relationships are within and between families.

According to our survey, conducted with information on plots, the owners, and farming methods, the number of plots totals 214, and there are an estimated 188 owners, most of whom have acquired the plots by inheritance.[4] As to access to land, 178 plots were inherited, mostly from fathers. Seven plots were inherited from mothers, and two from wives. Twenty-three plots were purchased.

Most owners belong to old families, probably because of the antiquity of this district (Table 48). In the palm tree orchard of Bir Nakhil, 36 owners belong to the Abd al-Hafiz family (*'a'ila*), 19 owners belong to the family of Jamal al-Din, and 12 to the family of Ababida. As mentioned in Chapters 5 and 6, the family of Abd al-Hafiz has been the most powerful family politically and economically, as the family of the umda (village chief). "Others" are mainly those who have moved into Rashda from Qasr, Tenida, Mut, Balat, and other villages.

Like the palm tree orchard (Bir Nakhil), the owners in the rice field (Bir Sheykh) are also concentrated in certain families. Those belonging to the

4 The majority of the owners have only one plot. However, there are 18 owners who have two plots, and three owners with three plots.

family Abd al-Hafiz own 17 plots, those of *Najjarin* own five plots,[5] and those of Jamal al-Din own four of 42 plots.

Water trading within a family can take multiple forms, because family is a flexible concept. For instance, a farmer can trade his water quota with his son or brother or relative because he is busy with non-farm activities, or because he is sick. It is also done between farmers of different families. This is most obvious in Bir Sheykh, as shown in Table 48. Seventeen of 42 owners belong to the Abd al-Hafiz family, but only seven cultivate. Most of their plots are contracted for share-cropping with farmers originally from Balat village. As explained in Chapter 5, those from outside villages such as Balat and Tenida villages came to settle in Rashda 100 years ago to work under the families that moved from Qalamun village; namely, the umda's family of Abd al-Hafiz. This historical tie is used in water trading.

However, it should be noted that transactions of water and labour resources may have an equalizing effect, but there is still great inequality in terms of access to water and land, which is largely because of asset inequality.

In fact, there is a considerable difference in the ownership of date palms. Of the owners of plots in the Bir Nakhil palm tree orchard, 66% have fewer than 10 date palms, and their plots account for only 30.1% of all the date palms—24.1% of the date palms belong to owners of plots with more than 40 date palms. The largest owner in terms of number of palms is A, who belongs to the family of AH. He has 240 date palms. The second and third largest owners, who have 84 and 57 trees respectively, are from small families. The latter owner with 57 trees lives in Mut city.

As for the Bir Sheykh rice field, there is also a large disparity between the owners in terms of area owned. The average area is 7.2 *feddan* per farmer, but the largest owner has a total of 28 *feddan*, followed by another farmer with 27 *feddan*.

Investment Well No. 51 District

Investment Well No. 50 is one of 10 "investment wells" created after 1996. As shown in Table 49, since the authorization of "investment wells" in 1995, many farmers have started to dig wells. Like other investment wells, No. 50 is located in an area somewhat far from the village (Map 26). No. 50 is located about 6 km north of the residential area in Rashda village.

5 On *najjarin*, see Chapter 7 and 10.

TABLE 48 *Number of owners in Bir Sheykh district by family* ('a'ila) (2011)

Family names	Rice field (Bir Sheykh)		Palm tree orchard (Bir Nakhil)
	Owners	Cultivators	Owners
AH	17	7	36
N	5	3	11
JD	4	4	23
Ab	3	1	12
F	2	1	12
AA			12
R	3		4
Kh	3	1	4
Other family originally from Qalamun	5	4	30
Originally from Balat village		4	1
Originally from Tenida village		4	5
Originally from Qasr, Mut			10
From other villages			3
Two or more families		5	
Unknown		8	5
Total	42	42	168

Note: The "other" 77 owners of the Bir Nakhil palm tree orchard include those belonging to families originally from Qalamun village or other villages.
Source: Information collected by informant in 2011.

The well is owned by seven colleagues working at an office. In 1996, Mr. A asked his colleagues to invest in this project, collected money from them, and received a licence from the Office of Investment (Maktab al-Istismar bil-Muhafaza) in Kharga and the Irrigation Office of the Ministry of Irrigation in Mut city to dig a well.[6] They paid the New Valley governorate to obtain 50 *feddan* of land for 50 Egyptian Pounds per *feddan* and spent 120,000 Egyptian Pounds digging a well 500 metres deep. It was dug by a local digger in Budkhulu village. Twenty *feddan* were added later, and in 2011, the area was 80 *feddan*.

6 It is forbidden for public servants to become beneficiaries of investment projects. For this reason, he made the application and received the licence in his wife's name.

TABLE 49 Investment wells in Rashda (2011)

No. Well	Year of creation	Depth	Method of water withdrawal	Area (*feddan*)
75	1996	550	Motor	200
73	1998	550	Artesian	200
74	2007	500	Motor	50
31	2000	350	Motor	50
50	1998	400	Motor	50
51	1996	480	Motor	50
52	2003	500	Motor	100
54	1003	500	Motor	50
53	2003	500	Pump	100
55	1998	500	Motor	100

Source: Information collected by informant.

Water Allocation

The area is irrigated by water drawn from a 24-hour well by diesel-powered pump during winter, and during the daytime. Water withdrawal from the artesian well at the time of digging was 30 *qirat* per day, but it decreased to 20 *qirat* after two years, then to seven *qirat* after another two years. Therefore, a diesel-powered motor was installed in 2000. Currently, water withdrawal is estimated to be around 20 *qirat*.

The rotation schedule is eight days with a pump that operates for 24 hours. Each plot receives two *amila* (day or night) every eight days. However, the actual rotation schedule is flexible and is discussed at a meeting held at the beginning of every season, and is decided according to the water demands of the crop. Although the water quota is negotiated between owners, they may exchange the water quota between the farmers; for instance, between the farmers who grow fodder crops or vegetables and those who have fallow land.

There is a Chief of Water among the seven initial owners, who is responsible for managing the water allocation. As compensation, he receives an extra two *amila* for the job. The cleaning of canals and other maintenance tasks are performed by the farmers themselves. All the owners must pay 1,000 Egyptian Pounds per month for operational costs such as electricity for running motors.

Cultivation System

The crop most favoured by the farmers seems to be alfalfa (*birsim hijazi*), because every owner grows it in one plot. The majority of farmers also cultivate palm trees and some fruit trees such as mango, lemon, or fig trees (Map 26). Wheat and Egyptian clover (*birsim baladi*) are also cultivated in winter.

Four owners farm by share-cropping (Table 50). In all cases, the owners take charge of expenditure including fertilizers, tractors, and seeds. Therefore, the owner receives two-thirds and the share-cropper one-third of production. If there are two share-croppers, as is the case for Owner No. 3, the plot is divided into two plots, and each share-cropper obtains one-sixth of production. Two owners rent a part of their plot for cash at the price of 4,500 Egyptian Pounds per year.

Interestingly, the owners are not from certain families. They are friends and from different families. Their common point is that they all have stable and relatively prestigious professions. For instance, Owner No. 1, who first had the idea, works in a public office in the city of Mut. Owner No. 2 is a primary school teacher, No. 3 is an Arabic teacher in a junior high school, No. 4 is an English teacher in a high school, No. 5 works as a secretary in a high school, and No. 7 works as a director in an electricity company.

Share-croppers are also from different families, but unlike the owners, who are originally from Qalamun village, they are mostly from other villages. Indeed, five of nine share-croppers are originally from Balat village.

Saada (Surface Spring) District

Saada is a district irrigated by a surface spring, located in the Duhus area south of Rashda in the direction of Mut city. Its well, named Amal, is one of the surface springs that increased rapidly in number during the 2000s. As shown in Map 27, there are many surface springs near the Saada district in the foothills area.

The Amal well was drilled in 2002, and it irrigates an area of 30 *feddan*. It is a shallow well with a depth of 112 metres, pumped with an electric motor pump of 20 horsepower. It cost 15,000 Egyptian Pounds to drill the well, in which the eight co-owners invested.

The land is developed government land, rented from the government. There is an electric motor pump that discharges 20 *qirat* (one *qirat* = approximately 100 cubic metres) of water every minute.[7] The total cultivable area is 30 *feddan*.

7 One *qirat* of water is defined differently according to region. In Rashda, one *qirat* is considered to be 100 cubic metres. One *qirat* may sometimes mean 125–250 litres of flow per minute (Sykes 1912: 726).

TABLE 50 *Cultivation system in "Investment well No. 50"* (2012)

Owner ID	Share of water	Cultivator (relation with owner)	Share between owner and cultivator	Area (*feddan*)	Number of plots	Family name of owner	Family name of cultivator
1	2 *amila*	Son	By family	10	7	R	R
2	2 *amila*	2 from same village	Each sharecropper one sixth and owner two third (in charge of expenditure)	10	9	F	Balat
							Balat
3	1 *amila*	2 from same village	Each sharecropper one sixth and owner two third (in charge of expenditure)			M	Qasr, Balat
	2/3 *amila*	1 from same village	Sharecropper one third and owner two third (in charge of expenditure)				Balat
	1/3 *amila* for palm date trees	Owner himself		10	10		
4	1 *amila* rented by cash	Brother	Rented to cultivator by cash			F	F
	1 *amila* sold	From same village	Sold			H	
	1 *amila* sold	From same village	Sold			F	
	1 *amila*	Owner himself		20	9		
5	2 *amila*	Husband of father's side aunt	Rented to cultivator by cash	10	5	N	Hijab
6	2 *amila*	Husband of the owner's wife's mother's side aunt	Sharecropper one thirdand owner two third (in charge of expenditure)	10	8	AH	N
7	2 *amila*	1 from same village	Sharecropper one third and owner two third (in charge of expenditure)	10	6	AH	Balat
Total	16 *amila*			80	54		

Note: One *amila* is one day (6 am to 6 pm) or one night (6 pm to 6 am).

TABLE 51 *Cultivated area in Saada irrigation district (2008)*

Plot ID	Cultivable land		Winter				Summer
	feddan	*qirat*	Wheat	Barley	Alfalfa	Vegetables	Alfalfa
1	5	0	1	0.5	3	0.5	3
2	5	0	1	0.5	3	0.5	3
3	5	0	1	0.5	3	0.5	3
4	5	0	1	0.5	3	0.5	3
5	5	0	1	0.5	3	0.5	3
6	1	8	0.5	4 *qirat*	0.5	4 *qirat*	0.5
7	1	8	0.5	4 *qirat*	0.5	4 *qirat*	0.5
8	1	8	0.5	4 *qirat*	0.5	4 *qirat*	0.5
9	1		0	0	1	0	1
Total	30	0	6.5	3	17.5	3	17.5

Source: Information collected by informant in December 2008.

In summer, evaporation is high, so only 17.5 *feddan* are used for cultivating alfalfa. The rest of the fields are kept as fallow land. In winter, wheat, barley, and vegetables are grown on 12.5 *feddan*, but currently the fields are covered exclusively by alfalfa (Table 51).

History of the Well

This project in this irrigation district was initiated by one co-owner (No. 1) (Table 52). The way in which partners were gathered shows the importance of relationships established in the workplace, family relationships, and friendship. The public administration, in this case the Local Unit, plays an especially important role in establishing relationships.

No. 1 saw an announcement by the Land Reclamation Fund offering the opportunity to drill a surface well and reclaim land. Therefore, he proposed the project to his former colleague at the Local Unit (No. 3), his relative (No. 5), and his friend living in Qalamun village (No. 6). Meanwhile, his former colleague (No. 3) proposed it to his friend (No. 2), who also proposed the project to his colleague (No. 4) at the Local Unit. A friend of No. 1 (No. 7) and one of No. 5 (No. 8) joined later, by buying water rights.

After gathering six people who were willing to share the project, No. 1 contacted a private drilling agency that specialized in drilling shallow wells. Then,

TABLE 52 *Co-owners of the Saada well (2008)*

Owner ID	Job	Relationship	Family name
1	Chief of a department in Local Unit	Initial idea	AA
2	Driver of Local Unit	Friend of No. 3	M
3	Specialist at family court	Ex-colleague of No. 1	AH
4	Service worker at Local Unit	Colleague of No. 2	AH
5	Agricultural engineer	Parent of No. 1	AA
6	Watchman of archeological site	Friend of No. 1	From Qalamun village
7	Farmer, livestock merchant	Friend of No. 1 (joined later)	AH
8	Technician at Local Unit	Friend of No. 5 (joined later)	From Mushia village

Source: Information collected by informant in December 2008.

he and his partners estimated the costs, searched for an area suitable for drilling a well and reclaiming land, started the administrative procedures, and made a plan to install a motor pump, to build a road, and to do other tasks.

Water Distribution

Water is distributed according to the share of investment by each co-owner in digging the spring. There is a total of 19 *amila*, and the owners decided that the watering rotation would be seven days. As mentioned in Chapter 10, one *amila* is one morning or afternoon, which corresponds to 12 hours. However, in summer, the operating hours of the motor pump are reduced to eight hours per *amila* (Table 53).

The watering schedule is decided as follows: shareholder No. 1 receives the water from 6 am on Saturday until 6 pm on Sunday, which is a total of 36 hours. No. 2, No. 3, No. 4, and No. 5 receive the water in a similar way. Owners No. 6 to No. 8 receive only 12 hours because their financial contribution was one-third that of owners No. 1 to No. 5.

In this way, the rotation is completed at 6 pm on Friday evening, which leaves one *amila* of 6 pm of Friday evening to 6 am on Saturday morning vacant. This one *amila* (Plot No. 9) is used for welfare. Thus, they fixed this *amila* as *waqf* and rented it to their shareholders or share-croppers for cash. The profit generated from this *amila* is donated to the Zakat Committee of Abu Bakr Mosque in Rashda.

TABLE 53 *Watering schedule for Ain Saada (Surface spring in Duhus area) (2008)*

Plot ID	Time	Day		Winter			Summer		
		First	Last	Amila	Hours	Area (*feddan*)	Amila	Hours	Area (*feddan*)
1	6am–18pm	Saturday	Sunday	3	36	5	3	24	3
2	6am–18pm	Sunday	Monday	3	36	5	3	24	3
3	18pm–6am	Monday	Tuesday	3	36	5	3	24	3
4	6am–18pm	Tuesday	Wednesday	3	36	5	3	24	3
5	18pm–6am	Wednesday	Thursday	3	36	5	3	24	3
6	6am–18pm	Thursday	Thursday	1	12	1.75	1	8	0.5
7	18pm–6am	Thursday	Friday	1	12	1.75	1	8	0.5
8	6am–18pm	Friday	Friday	1	12	1.75	1	8	0.5
9	18pm–6am	Friday	Saturday	1	12	1.75	1	8	1
	Total			19	228	32	19	152	17.5

Source: Information collected by informant in December, 2008.

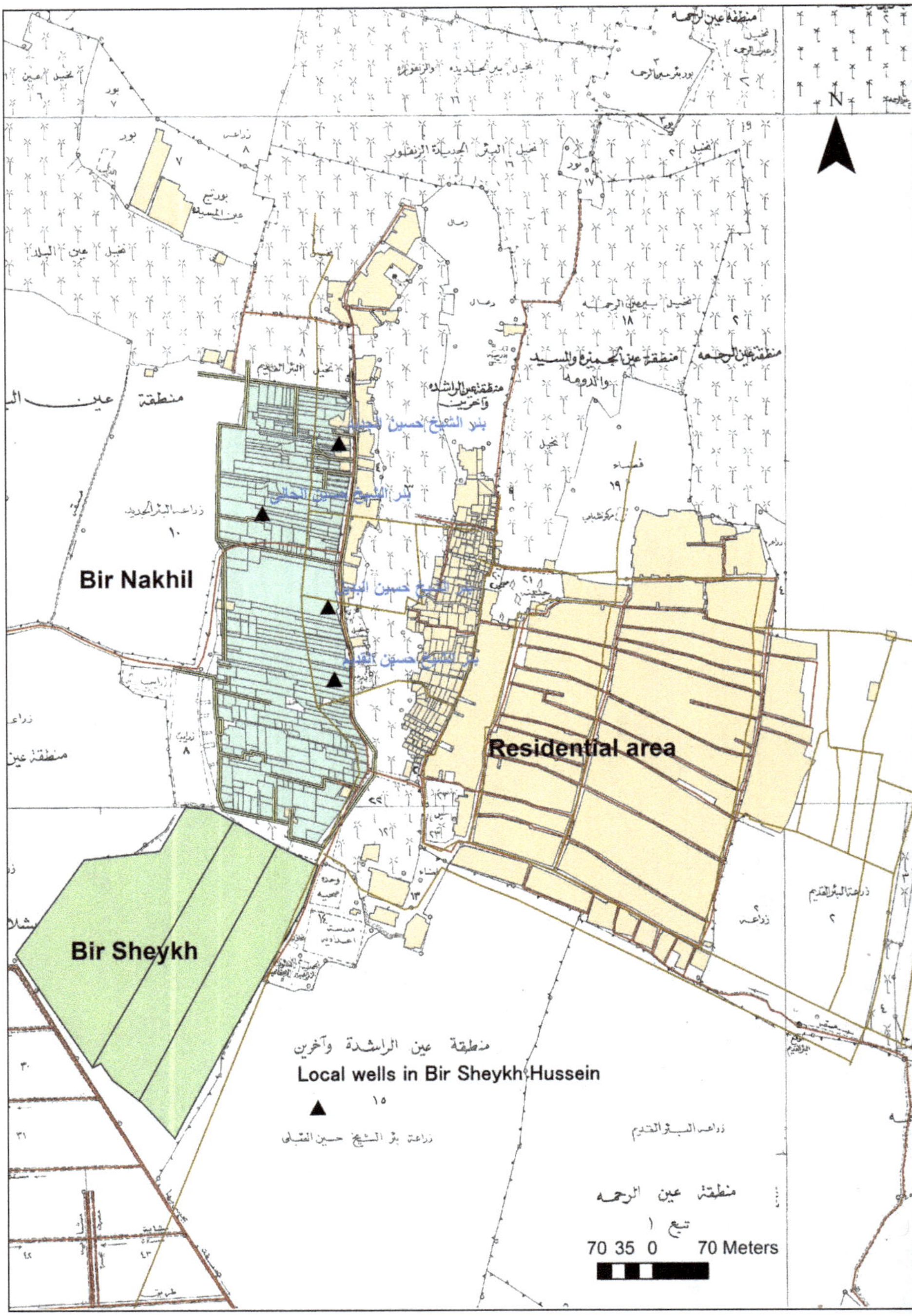

MAP 24 *Location of Bir Sheykh Husayn (Bir Nakhil and Bir Sheykh) irrigation district.* *Source*: Digital map developed based on fieldwork and information from an informant in 2011.

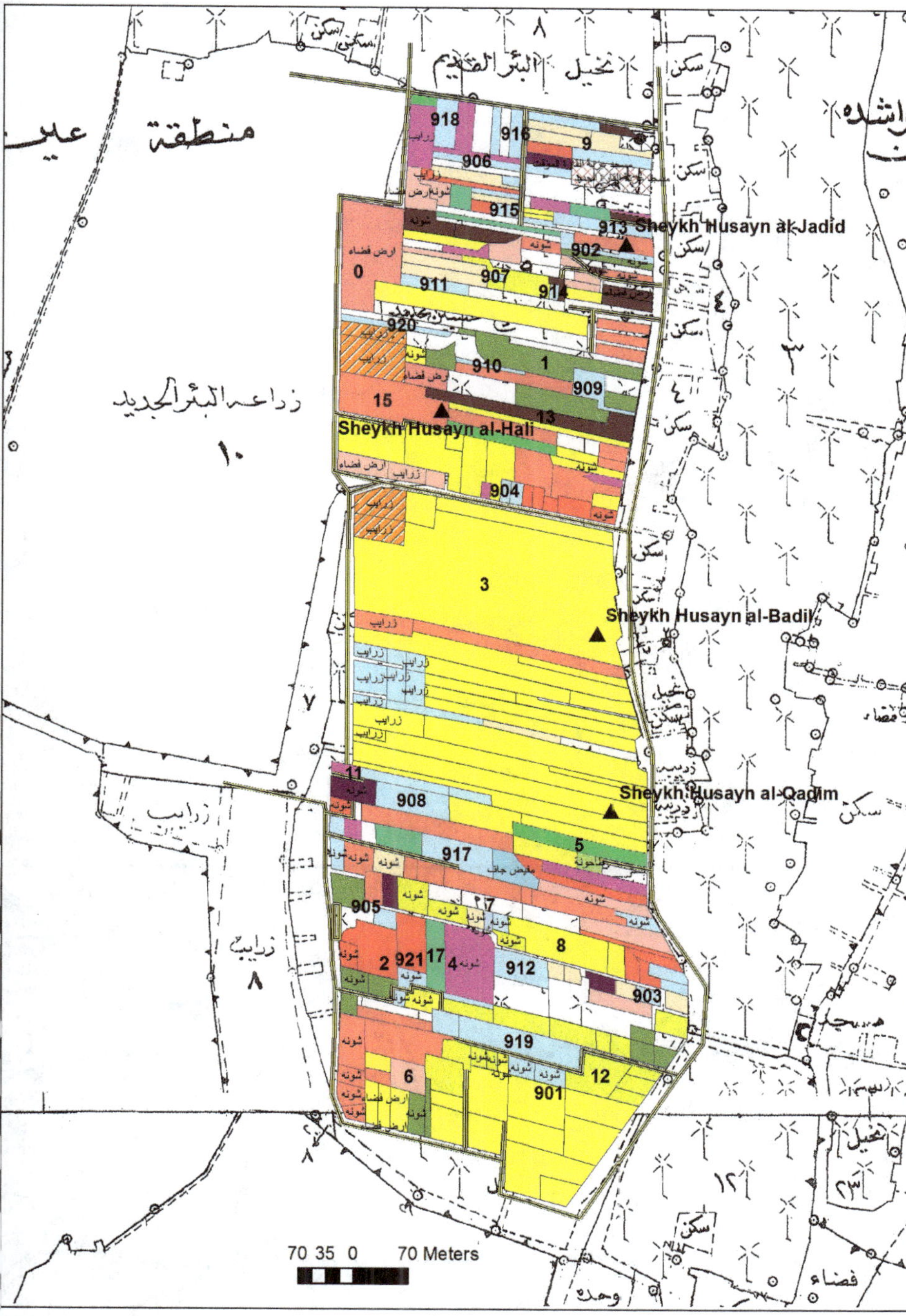

MAP 25 *Families owning plots in the wells in the Bir Nakhil palm tree orchard in the Bir Sheykh district. Codes refer to families. Source*: Digital map based on fieldwork and information collected by informant in 2010–2011.

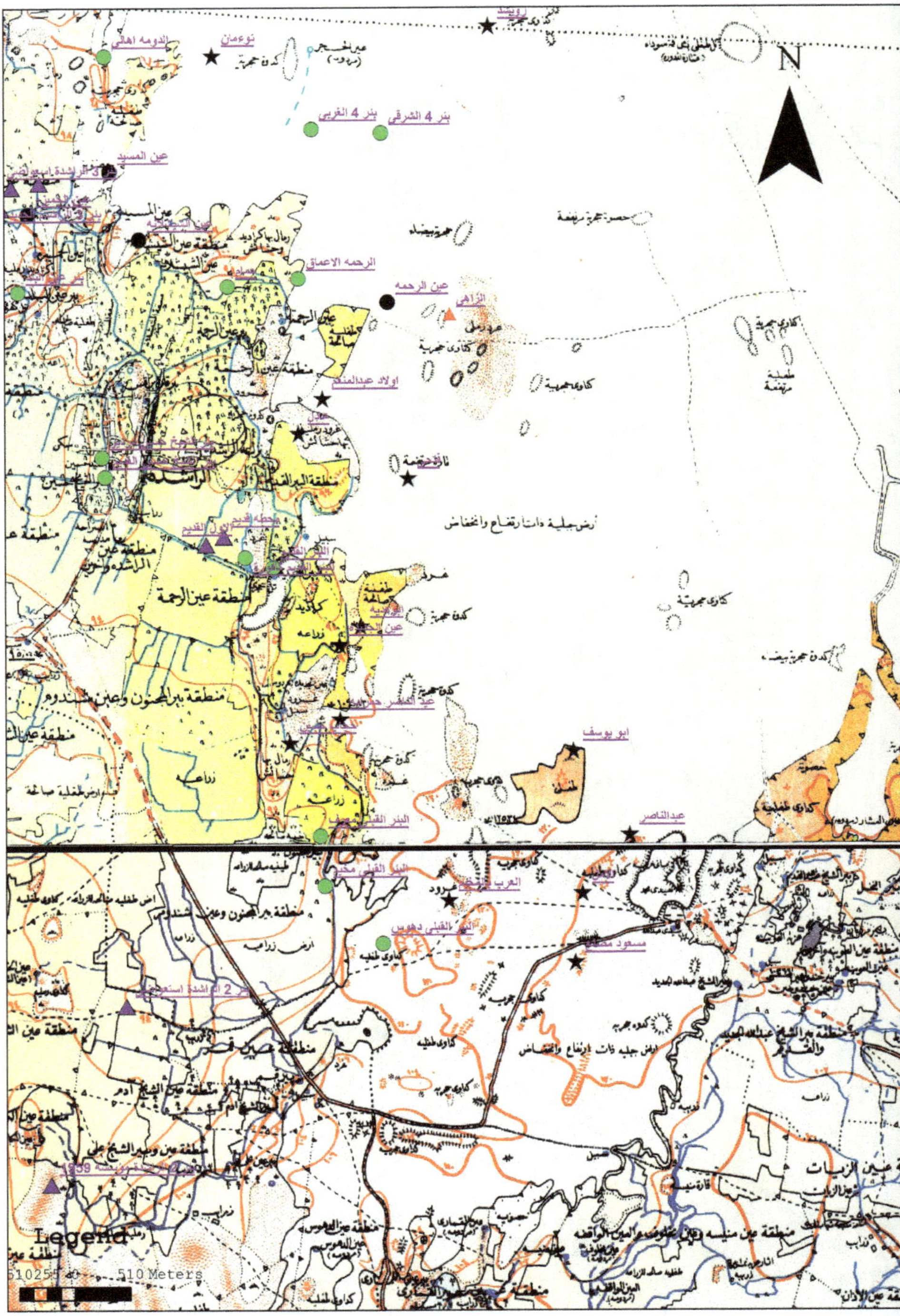

MAP 26 *Location of Investment Well No. 51 District. Source*: Digital map based on fieldwork and information collected by informant in 2012.

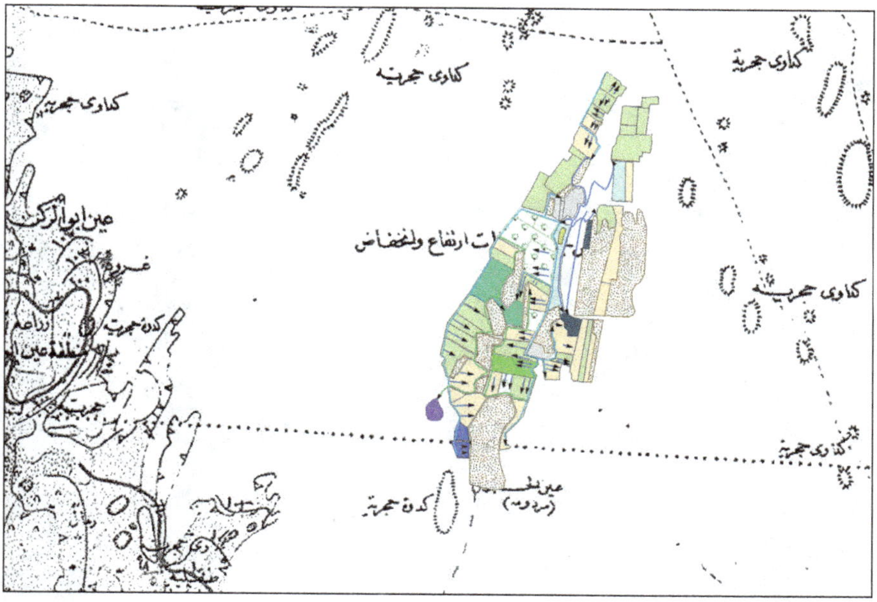

MAP 27 *Location of Saada district. Source*: Digital map based on fieldwork and information collected by informant in 2012.

PHOTO 39 *Bir Sheykh Husayn al-Hali (currently operating 'replacement well') in Bir Nakhil*

PHOTO 40 *Main street in Bir Nakhil*

PHOTO 41 Jiza (*rice gorge*) *of Bir Sheikh rice field*

PHOTO 42 *View of the field (alfalfa) in Saada district*

CHAPTER 13

Society of Well No. 3 Irrigation District

Chapter 12 discussed the irrigation and cultivation systems in irrigation districts irrigated by local wells and surface springs. This and the next chapter (Chapter 13 and 14) deal with an irrigation district irrigated by a government well, "Well No. 3 Irrigation District". The research in the two chapters is from an interdisciplinary perspective, using statistical and geographical information, including digital maps, satellite data, and the other qualitative information gathered during field surveys. This chapter discusses topics similar to those mentioned in Chapter 12 on irrigation districts irrigated by local wells and surface springs: its history, the irrigation management system and the crop cultivation system.

History of Well No. 3 Irrigation District

Well No. 3 Irrigation District is located in the north-west part of Rashda (Map 28). It is an agricultural field irrigated by five government wells—in practice, now four wells because one has dried up. The area was chosen for the well because it had been fallow land lying on the outskirts of land irrigated with the water from a local well.

The development of this district started as part of the New Valley Project by the government in the early 1960s.[1] In 1959, digging of the first well was authorized by the Land Reclamation Office in Mut city. It was contracted to, and drilled by, an Italian company to a depth of 500 metres with an initial discharge volume of 3,600 cubic metres per day.[2]

In the following year, 1962, the Egyptian government reclaimed the district for agricultural use and registered the land as *zimam*.[3] Then land preparation started, undertaken by a staff member at the Department of Irrigation in Mut city.[4] After experimental cultivation by a staff member at the Department of

1 On the New Valley Project, see Chapter 1.

2 The Italian company was nationalized after the 1952 revolution by Gamal Abdel Nasser and became REGWA (General Company for Research and Ground Water). The villagers still call the well No. 3 drilled by the Italian company "Italian well".

3 On *zimam* and *kharij al-zimam*, see Chapter 10.

4 The first harvest was in 1963, and the first fields in this district were ploughed in 1970.

 | DOI 10.1163/9789004317390_015

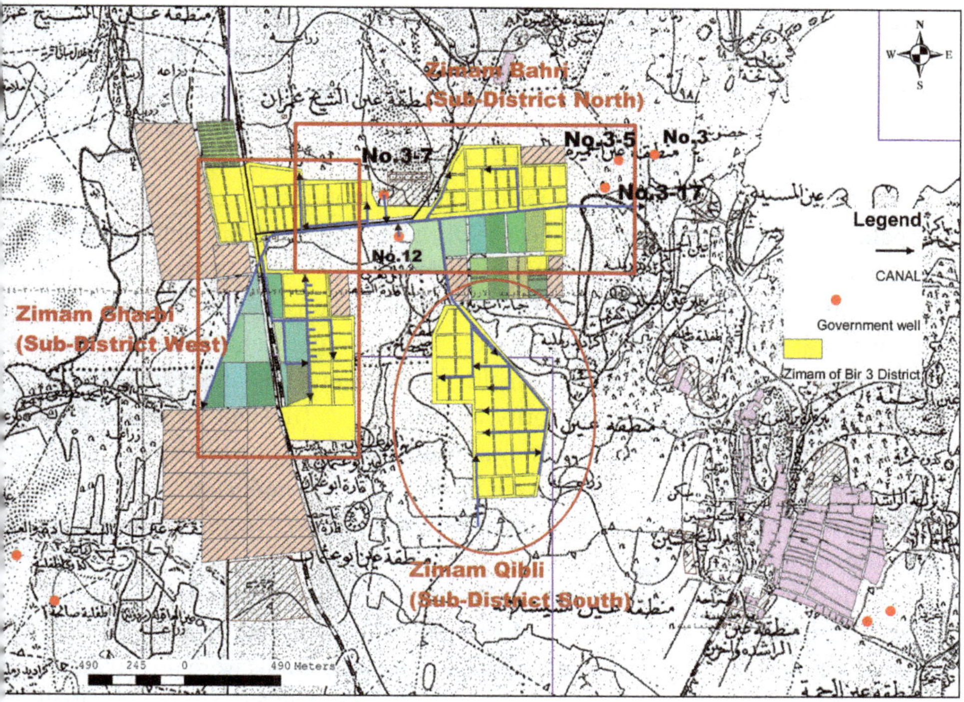

MAP 28 *Location of Well No. 3 Irrigation District. Source*: Kato et al. 2012: 250.

Irrigation of Ministry of Irrigation and Water Resources in 1963, the Land Reclamation Office of New Valley in 1964 sold a total of five *feddan* of land in the district to 34 farmers under 10-year repayment contracts.[5]

Officially, the area of this Well No. 3 Irrigation District is capped at 167 *feddan*, divided into three subdistricts (*zimam*) of North (Zimam Bahri), South (Zimam Qibli), and West (Zimam Gharbi). However, the actual area irrigated by the wells attached to this district is much larger, because of the complicated history of water use rights and the expansion of cultivated land outside the officially designated area (*zimam*).

A brief account of its historical development is given here. When the original land was reclaimed for agricultural use, there was already an orchard in the North subdistrict belonging to the land attached to the local artesian well, Ain

5 Experimental cultivation was undertaken by a staff member of the Department of Irrigation. This produced 800 *ardabb* of wheat, which were put in 150 sacks and sold to a villager at the price of 150 *qirsh* for each sack. The sacks were also distributed to the villagers who worked on the harvest.

al-Balad. This well dried up in 1950. Therefore, in the same year, the umda, who was the largest owner of water and land, made a claim to the police authority in Mut for the right to dig a new well. He made this claim to the police because Dakhla Oasis was under the authority of the police for administrative issues until the creation of the New Valley governorate.

The police authority offered 13,000 Egyptian pounds as compensation for palm trees that would be lost because of the drying up of the well. However, the umda refused this proposal after discussion with the farmers who shared the water from Ain al-Balad, and he insisted on the construction of a new well. For them, it was much more reasonable to have water rights than to receive cash that would quickly be gone. Finally, the umda and the farmers reached an agreement with the police authority, by receiving 15 (the sum of 10 and 5 for evaporation) out of 100 *qirat* per day of water discharged from Well No. 3, dug in 1959. Thus, the water rights were given to the farmers as a compromise between the government, which was willing to make an irrigation project, and the local farmers.

Another reason for the increase in the number of water rights holders is that wells are not sustainable; hence they need to be replaced by a new well in another place. In fact, from the end of the 1980s, four wells were dug to maintain the water discharge level: Well No. 3–5 in 1988, Well No. 3–7 in 1999, Well No. 3–12 in 2004,[6] and Well No. 3–17 in 2008 (Table 54). Meanwhile, the water discharge from Well No. 3 decreased, and ceased completely in 1995. It was an artesian well and could have continued to discharge water if a pump had been installed. However, instead of installing a pump, the Department of Irrigation subsequently drilled the new wells.

These new wells initially were artesian wells but currently have diesel-powered pumps. In Well No. 3–12, a diesel-powered pump with a discharge volume of 150 litres/minute was installed in 2006. Well No. 3–17 has been in operation since 2009. Well No. 3–7 is still an artesian well, and its current water discharge level is very low.[7] However, it is not possible to install a pump because of the trees covering the location of the well.

The important point is that the areas where the new wells were dug are privately owned land. Therefore, the owners of the land must be given a share of the water as compensation for letting the government use their lands. These cases concern the family of 10, who have an orchard in the neighbouring area,

6 Well No. 3–12 was dug in the land owned by a farmer. For this reason, this farmer has the right to obtain water from Well No. 3–12 for his land located outside the Well No. 3 Irrigation District.

7 Water discharge is around three *qirat* from Well No. 3–7, and it operates only in winter.

TABLE 54 *Wells in Well No. 3 Irrigation District (2008)*

Name of well	Year of digging	Depth of well (m)	Current water flow (cubic meters/24 hours)
Rashda No. 3	1959	500	Stopped (initially 3,600)
Rashda No. 3–5	1988	831	3,400
Rashda No. 3–7	1999	530	Nearly stopped (initially 3,413)
Rashda No. 3–12	2004	1046	2,300
Rashda No. 3–17	2008	739	2,900
Total quantity			8,600

Source: Information collected by informant.

and some farmers of Nakhil Mansura district in the neighbouring Budkhulu village.

A further land development has been the expansion of cultivated land in the northern and western parts outside of the official *zimam*. Despite expansion being officially frozen, much new land has been added. In the 1970s and 1980s, seven plots in the northern part outside of the official *zimam* have been purchased by 15 farmers. These plots were sold by auction and now are orchards of palm, mango, and other fruit trees. In addition, some plots were obtained through the act of farmers "putting their hands (*wadʿ al-yad*) on the earth", a culturally, but not legally, accepted method of acquiring land. Thus, the actual area of land attached to the wells in this district is higher than the official 167 *feddan*. This increase in quantity of land attached to the wells is an important factor of water distribution in this Well No. 3 District.

It is noteworthy that the expansion of cultivated land took place along with the digging of new wells by individual farmers, which became possible with the relaxation of underground water control by the government. The area to the left of the road leading to the West subdistrict is part of the district. The West subdistrict was initially located in this area but was moved to the current area around 1970 because of its high salinity. A surface spring was drilled in 2010, so that the land is now rented out to a farmer living nearby and cultivated.

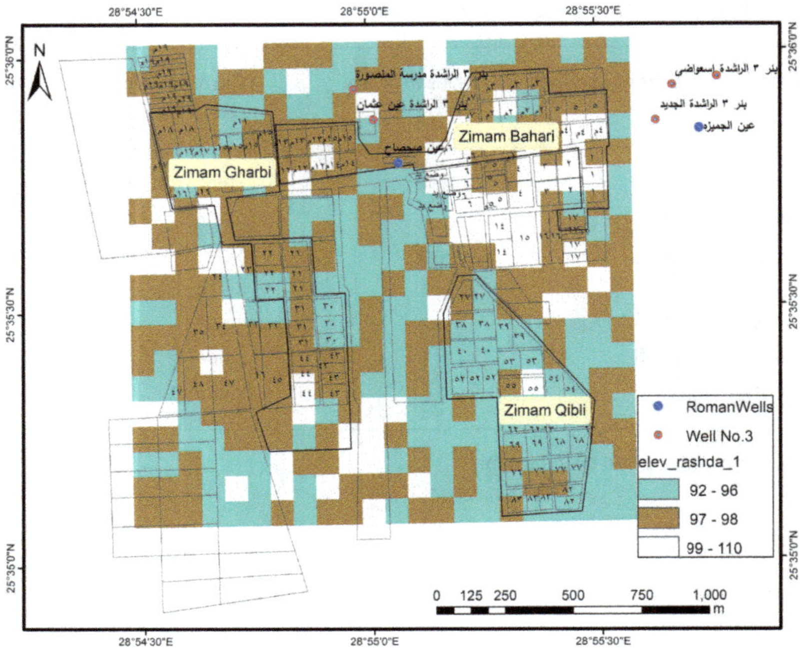

MAP 29 *Elevation levels of the three subdistricts. Source*: Kato et al. 2012: 260.

Land Use Change and Elevation[8]

The history of the Well No. 3 Irrigation District discussed above can be confirmed by observing the extension of cultivated lands around the district in satellite remote-sensing images, although this investigation only covers the period from 1972 to 1988 because of data limitations.[9]

One of the most important geographical features for irrigation is the elevation of the land. This determines the direction of water flow and influences the extension of cultivated land. Map 29 shows the elevation of the three subdistricts of the district estimated from Shuttle Radar Topography Mission (SRTM) data. It is noticeable that the elevation of the West and South subdistricts is lower than that of the North subdistrict.

---

8 This part is an extract from Kato et al. 2012, written by Dr. Salwa F. Elbeih.

9 Monitoring is based on the following three remote-sensing satellite images (Kato et al.): (A) Landsat MSS image (path 191, row 42) dated 12 November 1972 <http://edcsns17.cr.usgs.gov/EarthExplorer/>, (B) Landsat TM images (path 177, row 42) dated 19 December 1988 and 15 November 1984 <http://edcsns17.cr.usgs.gov/EarthExplorer/>, (C) SRTM elevation data dated February 2000 (www.jpl.nasa.gov/srtm).

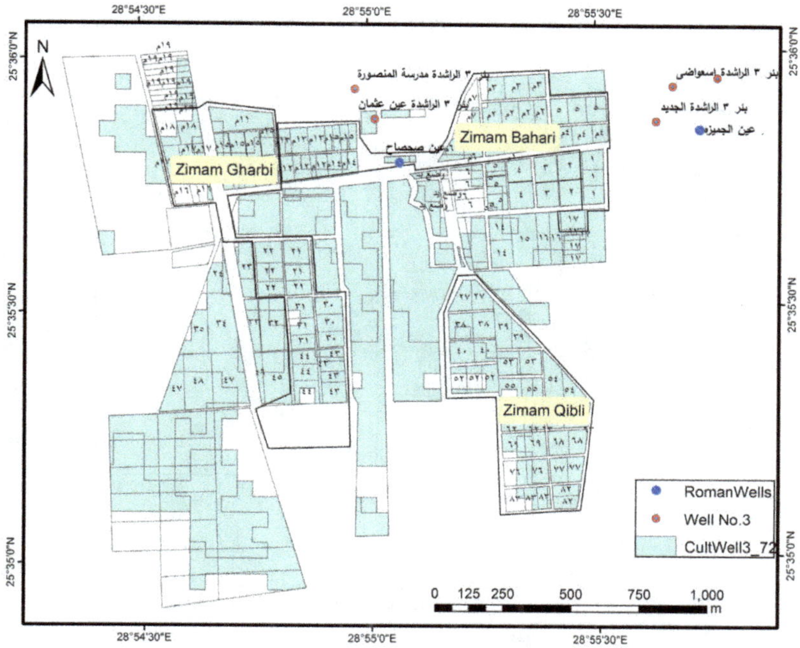

MAP 30.1 *Extension of cultivated lands around Well No. 3 Irrigation District (1972).* *Source*: Kato et al. 2012: 259.

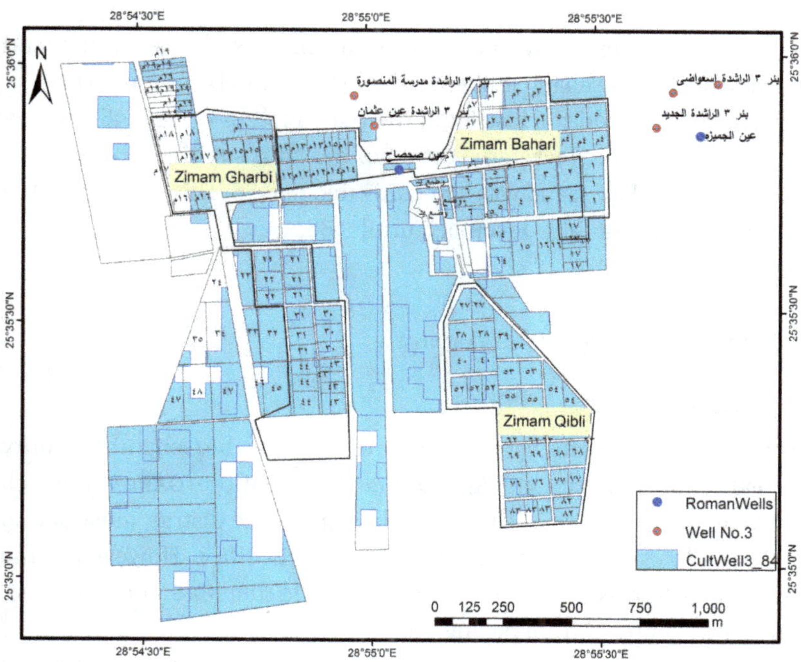

MAP 30.2 *Extension of cultivated lands around Well No. 3 Irrigation District (1984).* *Source*: Kato et al. 2012: 259.

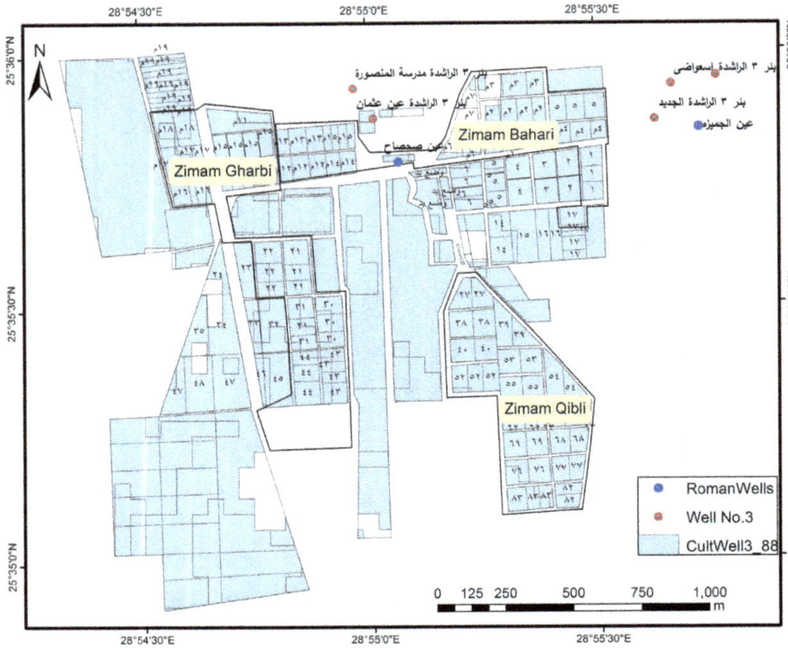

MAP 30.3 *Extension of cultivated lands around Well No. 3 Irrigation District (1988).*
Source: Kato et al. 2012: 259.

Map 30.1, 30.2, and 30.3 show the results of unsupervised classification and the extension of cultivated land in the Well No. 3 Irrigation District with respect to the Roman Springs and the district in the years 1972, 1984, and 1988. It may be observed that the north-western section of the West subdistrict was fallow land (*bur*) in 1972 and 1984, and was first cultivated in 1988. This means that the West subdistrict was not fully developed until 1988.

Irrigation Management System

Water Administration

Lands in Well No. 3 Irrigation District are administered by the General Authority for Reconstruction Projects and Agricultural Reclamation (GARPAR), which has its local office in Mut city. Tenants of the land in the district must pay 40 Egyptian Pounds every season (six months) to this office. However, if the amount of water flow is insufficient, and the land is therefore not cultivated, the tenants are exempted from paying the rent.

Water distribution, on the other hand, is under the control of the Irrigation Office. The Irrigation Office has fixed the water quantity at 22 m^3/day per

TABLE 55 *Water rights holders of the wells in Well No. 3 Irrigation District*

Water right holders		Area (*feddan*)	Water quantity (m³/day)	*qirat*/day
Original farmers	34 farmers who are beneficiaries of the governmental project in 1966	168	3,700	37.0
Owners of orchards	7 plots bought by 7 farmers in the 1980s, apiary belonging to Local Unit		850	8.5
Owners of palm trees in 'Bir Nakhil' (Ain al-Balad district)	2 owners of water rights of Ain Jumayza spring transferred to Well No. 3		300	3.0
	Includes farmers in Awina and Qalamun who own land in Ain al-Balad district		1,500	15.0
IO Family	Compensation for digging the well on the land of the IO Family		1,800	18.0
15 owners	Palm tree land inside district		326	3.3
Owners of palm tree district 'Nakhil Mansura'	Compensation for land nearby Well No. 3 District belonging to Budkhulu		423	4.2
Compensation for conducting the project	Local Unit		260	2.6
	AH Family			
Other 3 families	3 plots inside District		400	4.0
Drinking water	Izba Mansura		200	2.0
Total			9,759	97.6

Note: One *qirat* is 100 cubic metres.
Source: Information collected by informant in February 2015.

feddan in winter and 30 m³/day per *feddan* in summer, and it strictly controls the quantity of water discharge. The quantity is decided according to criteria such as soil quality and crop type. Because there are 167 *feddan* in the district, the farmers should receive a total of 3,674 m³/day in winter and 5,000 m³ /day in summer.

The remainder of the discharged water goes to other farmers who are water rights holders (Table 55). These include seven farmers who bought land in the district from the Department of Irrigation in the 1980s, the family of 10 on whose land the well was dug, farmers in the neighbouring palm tree land irrigated by a local well, the family of Abd al-Hafiz of the umda, who benefited from the water rights, and others. Thus, theoretically, the total quantity of water in winter should be 9,759 m^3 per day.

However, the actual amount of water discharge (8,600 m^3/day) is below this theoretical quantity, although the quantity of water that can be produced for the currently working wells (No. 3–5, No. 3–12, No. 3–17) is theoretically 17,200 m^3/24 hours (6,800 m_3, 4,600 m^3, 5,800 m^3, respectively). This is because the Irrigation Office reduced the quantity of water discharge by half by limiting operation from 24 hours to 12 hours (from 6 am to 6 pm). This reduction was intended to provide economic benefits such as reducing the cost of diesel fuel and hiring workers to operate the diesel pump, as well as reducing water consumption.

The result at the farm level is that the actual amount of water distributed in summer is 22 m^3/day per *feddan*, instead of 30 m^3/day as in winter. The winter growing season is from October to April. During this season, each *feddan* of agricultural land is irrigated with 22 m^3 of water once every 12 days. The summer growing season is from May to September. During this season, each *feddan* of agricultural land is irrigated once every 16 days. The reason for the longer irrigation interval during the summer season is the scarcity of water for irrigation. Because of the insufficient supply of water and the long irrigation interval during summer, the farmers cultivate only 60% of their land area. Therefore, many plots are left fallow.

Water Distribution

Map 31 shows the location of the wells and plots of the owners in Well No. 3 Irrigation District. The three wells—No. 3–5, No. 3–12, and No. 3–17—flow into lined well basins and subsequently into separate lined canals that allow the three streams of water to flow into the cultivated land. There are main canals, secondary canals, and tertiary canals that bring water to the three subdistricts. Secondary canals, which branch off the main line but remain separate from the agricultural fields, and tertiary canals, which lie next to farm plots, remain unlined.

At each point where a main canal connects with a secondary canal, or a secondary canal connects with a tertiary canal, there are junctures where the water flow is divided. These junctures consist of gates that remain open or closed, depending upon operational decisions of where the water should go.

Officially, decisions on how, when, and why to open these gates are in the hands of the water board and farmers' meeting. The board is composed of one

MAP 31 *Location of plots of the owners in Well No. 3 Irrigation District (2009). The numbers on the map identify each owner. Source*: Kato et al. 2012: 250.

employee of the Irrigation Office and two farmers appointed as managers of the wells. The meeting of farmers is held at the beginning of each agricultural season in May and October, and they make decisions regarding water distribution and the maintenance of wells and canals. This meeting is held in the presence of an engineer from the Department of Irrigation and should be held with all the farmers in the district in attendance. However, only a few farmers attend the meeting, either because the distribution and works are routine or because they are busy with non-farm work.

In the scheduling process, farmers are assigned a water quota every 12 days in winter and 16 days in summer. As pointed out in Chapter 10, the farmers conceive the quantity of irrigation water on a time rather than volumetric basis. The exact quantity of water that reaches each farm is known only in estimates. In the case of Well No. 3 Irrigation District, each cultivator of five *feddan* receives two *amila* and eight *qadam* (one *amila* equals 30 *qadam*) in one rotation as a general rule. The actual distribution of *amila* to each plot is shown in Table 56.

TABLE 56 *Size of plots in Well No. 3 Irrigation District (2008)*

Plot ID	South		North		West		Total area		Total quantity of water supplied in 3 plots (2007)	
	qirat	*feddan*	*qirat*	*feddan*	*qirat*	*feddan*	*qirat*	*feddan*	*amila*	*qadam*
1	23	1	15	1	14	1	4	5	2	8
2	13	1	3	2	2	1	18	4	1	15
3	2	2	11	1	15	1	4	5	2	8
4	2	2	11	1	8	1	21	4	2	4
5	6	2	4	1	12	1	22	4	2	4.5
6	13	1	21	1	2	1	12	4	1	29.5
7	21	1	14	1	16	1	3	5	2	7.5
8	22	1	16	1	13	1	3	5	2	7.5
9	12	1	5	1	20	1	13	4	2	.
10	13	1	17	1	15	1	21	4	2	4
11	13	1	19	1	5	1	13	4	2	.
12	2	2	14	1	16	1	8	5	2	10
13	4	2	11	1	16	1	7	5	2	7.5
14	20	1	18	1	3	1	17	4	2	2
15	7	2	12	1	13	1	8	5	2	10.5
16	0	2	13	1	15	1	4	5	2	8
17	0	2	23	1	5	1	4	5	2	3
18	2	2	14	1	5	1	21	4	2	4
19	2	2	13	1	5	2	20	5	2	6
20	1	2	11	1	12	1	0	5	2	3
21	1	2	0	1	18	1	19	4	2	3
22	2	2	14	1	2	1	18	4	1	28
23	20	1	16	1	14	1	2	5	2	6.5
24	18	1	17	1	13	1	0	5	2	6
25	16	1	13	1	18	1	23	4	2	6
26	14	1	13	1	8	1	11	4	1	29
27	14	1	13	1	15	1	18	4	2	3
28	5	2	4	1	5	1	14	4	2	3.5
29	5	2	2	1	5	1	12	4	2	.
30	6	1	21	1	15	1	18	4	1	23
31	5	1	2	2	14	1	21	4	1	19
32	9	1	16	1	15	1	16	4	2	1.5
33	8	1	15	1	17	1	16	4	2	1.5
34	8	1	10	1	19	1	13	4	1	15
Total	21	61	13	53	6	51	16	166	70	17

Notes: 1. One *amila* equals 30 *qadam*; 2. Most of the owners have died since the agricultural land was reclaimed in 1964; six owners are still living (as of 2009). Therefore, the land has actually been passed on to their sons.
Source: Kato et al. 2010: 30.

On the day when a farmer is to receive his water allotment, he personally goes to the field and removes the barriers (a movable metal dam or pile of mud) to let the water flow into his farm plot. Flood irrigation is the only method employed in the district, so each farmer lets water flow into his fields, letting water enter and soak the land until the next farmer closes the barrier and transfers the water flow to his own plot.[10]

Thus, the water distribution system seems to be very clear and simple. However, despite the officially determined irrigation schedule, the practical application of the schedule is constantly renegotiated at the farmer level. First, it should be remembered that the water is also distributed to the cultivated land outside the *zimam* that was previously irrigated by the water from the local well. It should be also recalled that more unofficial cultivated land has been added to the official land in the district since its establishment in the 1960s.

Second, the irrigation schedule distributes 37 *qirat* of water to 168 *feddan* of agricultural land per day, or two *amila* and eight *qadam* (one *amila* equals 30 *qadam*) for five *feddan* every 12 or 16 days, but farmers are free to distribute their hourly quantities of water in any manner that they choose. If a farmer has around four *feddan* of land to his name, he is given 24 hours of water every 12 or 16 days. Yet if he is a farmer that has "put his hands" on land and has additional land, he is not obligated to water those four *feddan*. It is up to the farmer to distribute his water. He may use his water according to his crop requirements (such as leaving some land empty, or combining water-intensive crops with non-water-intensive crops if he has additional plots outside the *zimam*). Thus, decisions outside of the schedule are commonplace.

This complex scheduling of water rotation is handled by one of the farmers. He is officially responsible for one of the wells in the district but is also informally responsible for the water rotation schedule—including all the water rights holders outside the district. He is a school teacher and has his plot inside this district and another plot in the neighbouring area. According to an informant, it is only this farmer who can make a complex schedule between the users of Well No. 3 government well *zimam* and those outside the *zimam*. As compensation for his jaala (*jaʿala*),[11] he receives one *amila*.

The maintenance works of the irrigation well and canals are theoretically collective because these works concern all the farmers sharing the same well. However, because the well is a government well, most of the tasks are performed by the Irrigation Office. An engineer from the Irrigation Department

10 On the irrigation system in Rashda, see Chapter 10.

11 On *jaala*, see Chapter 7.

visits the wells every month to check the quantity and quality of water. The wells are also cleaned at the time of this visit.

Cleaning of the secondary and tertiary canals, on the other hand, is the responsibility of farmers. Owners of plots must participate in cleaning the main canal every season. However, it is not unusual for them to be absent.[12]

Crop Rotation System

One of the characteristics of Well No. 3 Irrigation District is that, like other districts irrigated by government wells under the Land Reclamation scheme, the five *feddan* sold to each farmer are split into three plots, each in different subdistricts (the North, South, and West subdistricts).[13]

The allocation of agricultural land ownership was designed under the Land Reclamation scheme to ensure equality between the farmers. Each farmer received a plot near the well, a plot in the middle of the command area, and a plot far away from the well (see Map 22). This design was intended to reduce inequality in water flow between downstream and the upper course where water loss due to seepage and evaporation is worse than in the downstream fields. In this distribution system, all share both the benefits of having land near the water source and the disadvantages of having land far away from it.

Another reason for land ownership being split into three plots is related to the crop rotation system. The type of crops and their rotations are decided by the Directorate of the Ministry of Agriculture. Because rice cultivation requires permanent irrigation, and water flows for other crops must be controlled to avoid the overuse of water, crop rotation is necessary.

One of the three subdistricts must cultivate alfalfa (*birsim hijazi*) for three years, while two subdistricts participate in the crop rotation. For this reason, the crops are rotated between subdistricts every three years and six months.

No meeting is held to decide when and what kind of crops to cultivate, because the same cropping pattern is followed every year. However, in some circumstances that require the agreement of cultivators, such as the prohibition of rice cultivation, the Agricultural Co-operative assumes the task of informing and directing the cultivators.

Before the prohibition of rice cultivation, the government regulated the cultivation of paddy rice, depending on the amount of water available for

12 On canal cleaning, see Chapter 8.

13 The actual plot area is 4.5 *feddan*, because 0.5 *feddan* is used for drainage of canals and streets.

Table [illegible] *Crop rotation in Well No. 3 Irrigation District (1999–2012)*

Year	North subdistrict		South subdistrict		West subdistrict	
	Summer	Winter	Summer	Winter	Summer	Winter
1999	Fallow land	Wheat and vegetables	Alfalfa and fallow land	Alfalfa, partially vegetables	Fallow land and rice	Egyptian clover
2000	Rice	Egyptian clover	Alfalfa and fallow land	Alfalfa	Fallow land	Wheat
2001	Fallow land	Wheat	Rice	Egyptian clover	Alfalfa	Alfalfa
2002	Rice	Egyptian clover	Fallow land	Wheat	Alfalfa	Alfalfa
2003	Fallow land	Wheat	Rice	Egyptian clover	Alfalfa	Alfalfa
2004	Alfalfa	Alfalfa	Rice	Egyptian clover	Fallow land	Wheat
2005	Alfalfa	Alfalfa	Fallow land	Wheat	Rice	Egyptian clover
2006	Alfalfa	Alfalfa	Rice	Egyptian clover	Fallow land	Wheat
2007	Fallow land	Wheat	Alfalfa	Alfalfa	Maize/sorghum (instead of rice)	Egyptian clover
2008	Maize/sorghum (instead of rice)	Egyptian clover	Alfalfa	Alfalfa	Fallow land	Wheat
2009	Fallow land	Whcat	Alfalfa	Alfalfa	Maize/surghum (instead of rice)	Egyptian clover
2010	Maize/sorghum (instead of rice)	Egyptian clover	Fallow land	Wheat	Alfalfa	Alfalfa
2011	Fallow land	Wheat	Maize/sorghum (instead of rice)	Egyptian clover	Alfalfa	Alfalfa
2012	Rice	Egyptian clover	Fallow land	Wheat	Alfalfa	Alfalfa

Note: In 2009, the West subdistrict was designated to be mainly used for rice in summer and Egyptian clover in winter. However, because of a water shortage, maize was cultivated instead of rice in summer.
Source: Information collected by informant in 2012.

irrigation. Every year, the managers of a well and the farmers held discussions to determine how much agricultural land could be planted with paddy rice. They took into consideration the flow of underground water and the results of pumping tests at the well, and they anticipated the irrigation conditions for that year to determine the area to be cultivated.

Table 57 shows the crop rotation from 1999 to 2012. An important point is that when rice was grown in summer, Egyptian clover (*birsim baladi*) must be cultivated in winter to let the soil rest. However, because of water shortages in recent years, rice was prohibited so that the land was used either as fallow land or for Egyptian clover cultivation in the rice season.[14]

PHOTO 43 *Well No. 3–12*

14 Few fields had been planted with paddy rice in the summer of 2009, because of the scarcity of water for irrigation. Many fields had been left fallow. Paddy rice in this field was being cultivated illegally. In such cases, farmers are fined. For paddy rice cultivation, continuous irrigation is necessary during a growing period. Therefore, one division was narrowed from its normal width, and the discharge was reduced to ensure a continuous supply of water to the paddy field.

PHOTO 44 *Main canal in Well No. 3 South subdistrict*

PHOTO 45 *Rice fields in Well No. 3 Irrigation District*

PHOTO 46 *Fallow land in Well No. 3 Irrigation District (September)*

CHAPTER 14

Social Relations of Well No. 3 Irrigation District

Since Well No. 3 Irrigation District is an irrigation district irrigated by a government well, it is considered the best research field for a case study on the sustainability of a society living under conditions of limited resources, especially water, with reference to the relationship of the three concepts mentioned in the Introduction: the uncertainty of the water supply, dependence on the political regime, and the rational behavior of individuals. This chapter deals with the irrigation and cultivation systems, focusing on the social relations in the District: its formation by the government, the family system as an institution for landholding and management, and the agricultural co-operation between farmers.

An Irrigation District Formed by Government Policy

Family Relationships in the District

As described in Chapter 13, Well No. 3 Irrigation District is irrigated by a government well and consists of plots allotted to farmers by the government. Water management is administered by the Irrigation Office, and the crops to be cultivated are decided by the Directorate of the Ministry of Agriculture. Thus, the commitment of the state distinguishes the Well No. 3 Irrigation District from other districts irrigated by local wells, investment wells, and surface springs.

The idea of this government-run project was socialist and based on land reform. For this reason, the Socialist Committee that took over the leadership of the village administration at the time of establishment of the government projects in the 1960s avoided choosing candidates from the umda's family, although some of his family members met the criteria. At that time, it was thought to be shameful for the umda's family to participate in a project for the poor.

In fact, the 34 beneficiaries who obtained land in 1964 were chosen by the village Socialist Committee because they satisfied the following criteria: 1. they had completed seven years of military service, especially in the Yemen War, 2. they had worked as waged agricultural workers in projects run by the Land Reclamation Authority, 3. they had large families and many dependents, and 4. they were landless workers.

 | DOI 10.1163/9789004317390_016

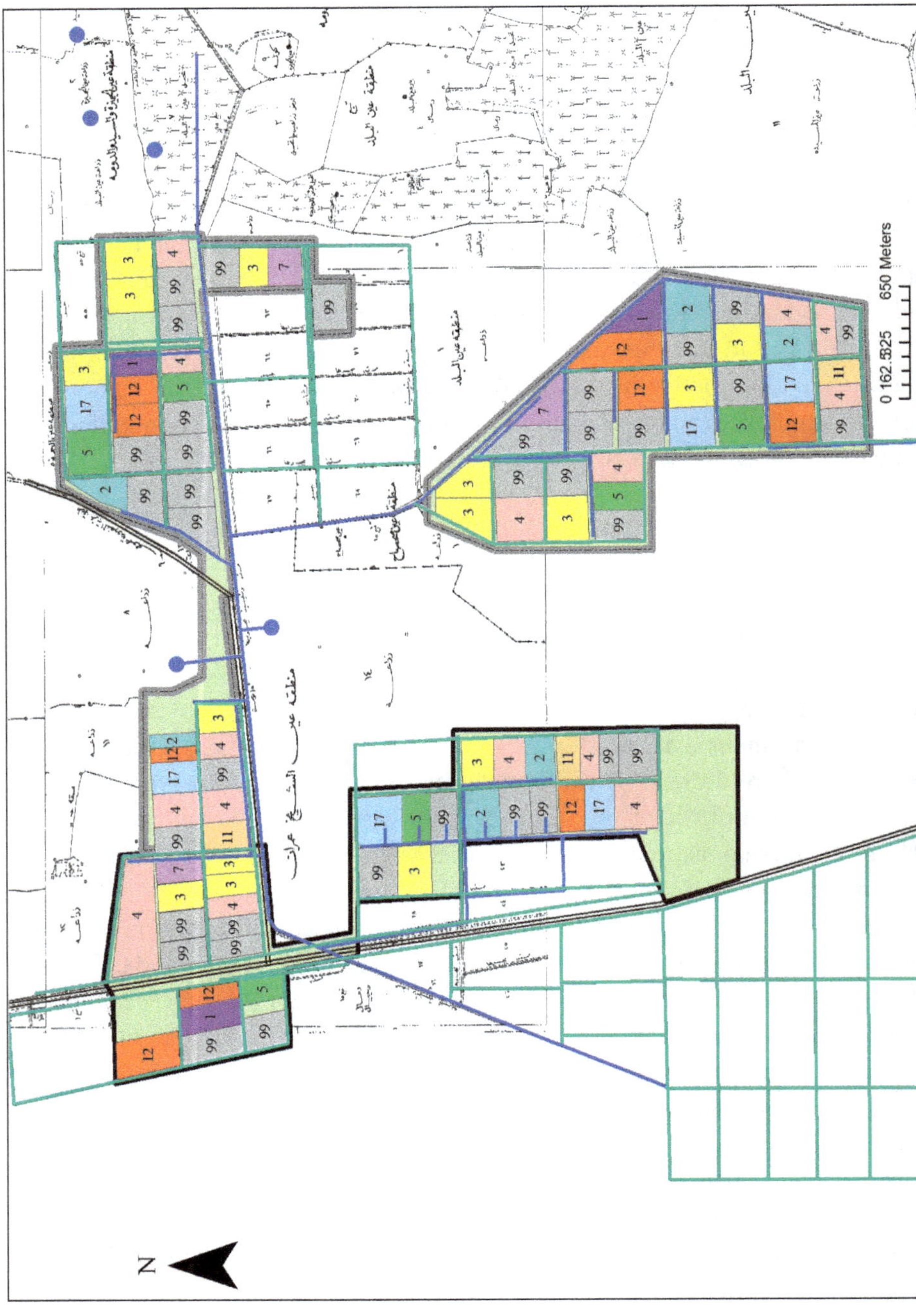

MAP 32 *Owners by family in Well No. 3 Irrigation District. Source:* Digital map based on the information collected by informant in 2009.

Therefore, a feature of Well No. 3 Irrigation District is the absence of traditional social relationships based on family. In fact, in contrast to the districts irrigated by the local wells, few farmers belong to the old families in this district. As was shown by the case of Bir Sheykh district in Chapter 12, there is a wide disparity in the local well districts between the farmers sharing the same well and the form of water and land accumulation by the umda's family. In the case of the Well No. 3 District, 15 farmers who were the initial owners belonged to the old families, of whom only five belonged to the umda's family. This contrasts with the Bir Sheykh district, where 40% of the owners of rice fields and 21% of the palm tree orchardists belong to the umda's family.

Map 32 shows evidence of the absence of family relationships in this district. The grey plots are owned by farmers originally from outside the area. The coloured plots that are not grey belong to the old families. From this map, it is clear that the lands have not been accumulated by certain old families, unlike the Bir Sheykh palm tree orchards.

In addition, the 34 initial owners included 12 men who were originally from other villages such as Balat, Tenida, and Qasr. This also contrasts with the districts irrigated by local wells. In the case of Bir Sheykh District in Chapter 12, only five of the owners originally from other villages were among the 42 owners in the rice field, and there were 19 from other villages among the 168 owners of the palm tree orchards.

Musharaka *as a Farming Strategy*

However, the strong commitment of the government and the absence of family relationships do not mean the absence of co-operation between the farmers. The Well No. 3 Irrigation District is no exception to communal practices, as long as the farmers rely on the limited underground water that must be shared between them. In fact, as mentioned in Chapter 13, water management is administered not only by the Irrigation Office but also by one of the farmers who co-ordinates the water distribution within the *zimam* according to the jaala system.[1]

The key to understanding the farmers' strategy of co-operation with limited resources is *musharaka*. *Musharaka* literally means "co-operation" or "partnership". *Musharaka* takes diverse forms according to the circumstances, as is

1 Currently, there are no other farmers who work by the jaala system, besides the one who is in charge of water distribution co-ordination. In the initial period in 1960s, there were other kinds of jaala such as jaala for land preparation, canal cleaning. However, these jaala disappeared because the main canal was renovated in concrete, and manual labourers on the land were replaced by tractors, according to the informant.

understood from the diverse usage of the word *musharaka* in Rashda. *Musharaka* in cultivation takes two main forms. The first type is *musharaka* at the land management level. It may take the form of share-cropping or of family collective management. In Rashda, both forms of management are called *musharaka*. Another type is *musharaka* on occasions such as harvesting, which is the type described in Chapter 8. It is basically of a reciprocal nature, as in the "yui" practised by Japanese farmers as reciprocal labour exchange.

Family as an Institution of Landholding and Management

Land Fragmentation

Family is at the core of communal relationships in this district as well as other irrigation districts in Rashda. As long as resources are limited and distribution is undertaken according to Islamic inheritance law, which gives equal rights to sons (and to some extent to daughters), a farmer's rational choice is to co-operate within his family.

Of course, this is not the case if the water and land are transacted in a market system. However, in the case of Well No. 3 District, transaction is rare—probably because the district is administered as a government reclamation project. In addition, the farmers would prefer to keep the land, despite the land fragmentation, for household consumption. Having land of their own would improve the quality of the food consumption.

In fact, between 1964 (when the district was founded) and 2015, only four initial owners sold their plots, because of the cash needed for building a house, sickness, and other reasons. Except for two plots still owned by the initial owners, the original owners have passed away, and their land has been inherited by their children.

The rarity of land transactions means that the land becomes divided into ever smaller pieces because of Islamic inheritance law. Indeed, the number of land owners (or those who have inheritance rights) is estimated to be 257 people, and the number of plots has grown from the original 34 to 95. Among the plots owned by the 34 initial owners, only two are still held by the initial owners. The remainder of the 32 initial owners have passed away, and their plots have been inherited and divided into 95 plots by their sons and daughters.

As a result of Islamic split inheritance, whereas each owner once held an average of five *feddan* of the total area of 166 *feddan* and 16 *qirat*, the estimated current ownership is an average of 0.7 *feddan* per owner. The average area of land owned is becoming smaller and smaller.

TABLE 58 *Profile of land-owning households in Well No. 3 Irrigation District (2009)*

		Cultivating	Not cultivating
Age	Household head	53.5	51.9
	Household members	37.7	37.7
Average year of education	Household head	5.6	6.9
	Household members	7.3	7.8
Household size	Number of persons	4.4	3.7
Labour participation rate	%	3.5	3.1
Number	Household	34	34
	Household members	151	126

Source: Household survey 2009, information collected through fieldwork in 2010.

Household Allocation of Labour

Another motivator of co-operation between farmers is the necessity of household labour allocation in the context of non-farming sector development. In this section, we see the profiles of the households that have land among those surveyed in 2009 (see Chapter 11).[2] Among the 715 households surveyed in 2009, there were 68 that have land in the Well No. 3 Irrigation District. Of these households, 90.0% report that they receive income from agriculture, and 54.0% cultivate their land. This means that they let someone else cultivate their land, in return for a portion of the produce.

Household engagement in agriculture may be influenced by the household's age composition, labour force, educational level, and income level. Among these factors, labour availability and educational level are correlated with households' decisions whether to cultivate. In fact, the result of comparison between the households that cultivate land and those that do not among the 68 households owning land in the district reveals that the households that cultivate have lower average educational levels and a higher labour participation rate (Table 58). Moreover, they have fewer members engaged in non-farming waged jobs (Table 59). Thus, the households with available labour allocate it to cultivation, and those without sufficient labour may not cultivate.

2 With the help of an informant, we identified the households in the 2009 survey that owned land in the Well No. 3 Irrigation District, and then identified their plots.

TABLE 59 *Work status of household members aged 15 or older of the land-owning households in Well No. 3 Irrigation District (number of people) (2009)*

	Cultivating	Not cultivating	Total
Self-employed without workers	13	3	16
Self-employed with worker	4	7	11
Unpaid worker	30	11	41
Waged worker	37	47	84
Unemployed	5	7	12
Not working	34	29	63
Total	123	104	227

Source: Household survey 2009, information collected through fieldwork in 2010.

Musharaka *as a Land Management Strategy*

One way of responding to land fragmentation and labour scarcity is not to divide the land. Inheritance does not mean that the management of plots is divided between the inheritors. In fact, although ownership is clearly divided for all of the plots of 32 initial owners, only one is completely divided at the cultivation management level (Table 60). Eight plots were inherited by the original owners' children and were divided legally but not in practice. Twenty three plots were inherited and partially divided at the cultivation level. "Partially divided" here means that some inheritors manage cultivation together, even though the land was split between inheritors, while others do so individually.

This kind of cultivation strategy is observed not only in this district but also in other irrigation districts, because Islamic inheritance law and labour scarcity problems apply to all farmers in general. Among the 23 initial plots, which were divided into 79 plots, 59 are managed either by the inheritor-owner himself (41 plots) or by share-cropping by half shares with a villager (18 plots). In the cases of 28 plots, one inheritor-owner is in charge of cultivating his and other heirs' (such as mother's, sister's, or brother's) share of the inherited land. The plot rented out for cash is only one case.

A typical example of partially divided plots is as follows. After the initial owner passed away in 1983, an area of five *feddan* was split between his five children (two sons and three daughters) in 1990, according to Islamic law, which made the shares as follows: 2/7 for each of the two sons and 1/7 for each of the three daughters. All these inheritors managed their shared individually,

TABLE 60 *Cultivation management of inherited land (2015)*

		Initial plot number	Current plot number
Completely divided	Individual management		5
	One person in charge of his own and other inheritors' shares of land		1
	Total	1	6
Partially divided	Individual management		41
	Share-cropping by half		18
	One person in charge of his own and other inheritors' shares of land		28
	Other		1
	Total	23	79
Not divided		8	8
Not divided yet		2	2
Total		34	95

Source: Information collected by informant in February 2015.

except for one daughter who ceded the management of her part to her brother. Therefore, one brother cultivates 2.1 *feddan*, which is approximately 3/7 of the initial land area on behalf of himself and his sister.

A similar example is that of the children of an initial owner of the plot, who passed away in 2004. He had three sons and three daughters, so his land was divided between the sons, who each received three *feddan*, and the daughters, who each received half a *feddan*. The three brothers do their cultivation together, while the three sisters ceded the management of their land to a villager in return for half of the crop.

As for eight examples in which the plot is not divided, they are inherited and legally divided, but the cultivation is done collectively as one plot. Management of these cases can be classified into three patterns (Table 61). One is that the land is managed by another villager in return for half of the crop. The second pattern is that one or two inheritors (sons) take responsibility for cultivation on behalf of all the inheritors. The third pattern is that all the inheritors cultivate individually but harvest together.

According to an informant, there are many different reasons for a family to choose its method of management. It may be because the children who

TABLE 61 *Plots that are inherited but managed as musharaka (2015)*

	Number of inheritors	Cultivators	Mode of cultivation
Case 1	6 (wife, 2 sons, 3 daughters)	1 villager	share-cropping by 1/2 done by a villager
Case 2	13 (wife, 5 sons, 7 daughters)	1 son	one inheritor responsible for cultivation
Case 3	7 (1 wife, 1 son, 5 daughters)	1 villager	share-cropping by 1/2 done by a villager
Case 4	6 (3 sons, 3 daughters)	2 sons & 1 daughter's husband	share-cropping by 1/2 done by 3 inheritors
Case 5	9 (5 sons, 4 daughters)	2 sons	two inheritors responsible for cultivation
Case 6	8 (4 sons, 4 daughters)	1 grandchild	one son of inheritor responsible for cultivation
Case 7	14 (wife, 6 sons, 7 daughters)	2 sons	family cultivation (two inheritors responsible for cultivation) The products are consumed together.
Case 8	5 (5 sons)	1 son & 1 villager	share-cropping by 1/2 done by an inheritor and a villager

Source: Information collected by informant in February 2015.

inherited are not interested in agriculture, or are busy with non-farm activities, or in the case of daughters, they do no agricultural work by themselves because of customs. This may be because the land inherited is too small for individual agricultural holdings.

Four cases chosen from the above Table 61 describe the reasons why the inheritors share cultivation.

Case 1 Share-cropping by Half with a Villager

After the initial owner passed away in 2012, the land was inherited by his wife, two sons, and three daughters. However, the land was not divided, because they preferred to share-crop the land with a villager. The inputs, including seeds, fertilizers, and chemicals, were divided equally between the share-cropper and six inheritors.

The inheritors preferred share-cropping because they are old, and their children have non-farm jobs. It is also because they have had a good relationship with the share-cropper since 1970.

Case 2 Share-cropping by Half with a Brother

After the initial owner passed away in 1986, the land was finally divided between his wife, five sons, and seven daughters. However, the cultivation is managed by one of the sons. In addition to his part of the land, he manages the land inherited by his mother, brothers, and sisters by share-cropping by half. His brothers preferred to give him the land for share-cropping, because two of them live in Cairo; others are busy with cultivation in an area irrigated by another well.

Case 3 Share-cropping by Half with a Villager

After the initial owner passed away in 2003, the land was formally inherited and divided between his wife, one son, and five daughters. However, the land was not divided, because they preferred to share-crop by half with a villager.

They preferred share-cropping because the son was busy with his non-farm activities, and the other inheritors are all women. According to them, women would not work in the fields.

Case 7 Family Collective Cultivation

Case 7 is a special case. The initial owner passed away in 1998, but the land has not yet been divided. Fourteen people hold inheritance rights, including the wife, six sons, and seven daughters. Two sons are responsible for the management of the land in Well No. 3 Irrigation District. Other sons are responsible for the other plots that the initial owner owned in other places. After harvesting, they divide the produce between the sons and daughters.

Agricultural Co-operation between Farmers

Table 62 shows the farmers with whom each farmer co-operates as *musharaka* in the district. Map 33 shows the farmers associated with six other farmers in *musharaka*. As described in Chapter 7, *musharaka* here indicates co-operation in exchange for agricultural works, such as harvesting, planting rice, reparation of roads and canals, etc. From Table 62, it is apparent that there is no systematic co-operation within the same irrigation district. The co-operation is often achieved by mobilizing neighbouring farmers, as Map 33 shows, but there are many exceptions for farmers who are not neighbours. Rather, co-operation occurs on occasions that require manual work, using social networks of a different nature.

Musharaka is often done for wheat and rice, because these crops demand considerable manual work. Of course little manual work is required in modern

TABLE 62 *Musharaka between the farmers in Well No. 3 Irrigation District (2008)*

ID	1	2	3	4	5	6	7	8	9	10	11	12	13	14	15	16	17	18	19	20	21	22	23	24	25	26	27	28	29	30	31	32	33	34
1		●			●			●			●										●	●							●				1	
2				●	●	●		●	●	●	●	●	●	●		●		●	●	●	●	●		●	●		●		●	●			●	
3				●			●	●			●	●		●	●			●	●	●	●	●			●		●		●	●			●	●
4					●	●		●			●	●	●	●				●	●	●	●	●	●		●	●	●		●	●				
5						●	●	●			●	●		●				●	●	●	●	●			●		●		●	●				
6														●							●		●									●		
7								●	●																					●				
8									●		●						●									●								
9											●						●									●								
10											●							●	●	●														●
11																					●	●												
12													●						●									●	●	●				
13																							●									●		
14															●																	●		
15																												●					●	●
16																	●					●									●			
17																										●							●	
18																			●															●
19																				●														
20																					●					●	●							
21																						●	●											
22																							●								●			
23																																		
24																									●									
25																																		
26																											●							
27																															●			
28																													●					
29																																		
30																																		
31																																●		
32																																		
33																																		●
34																																		

Note: A black dot indicates co-operation (*musharaka*) in agricultural tasks such as harvesting, planting rice, and repairing roads and canals.
Source: Information collected by informant in 2008.

times. In the past, transportation of fertilizer made from cattle manure compost was a heavy job that needed four days to transport it using 10 to 12 donkeys. However, no such job is necessary following the introduction of chemical fertilizer and trucks. However, there is still manual work that must be done.

For instance, in rice cultivation of one or two *feddan*, harvesting requires three to five workers, threshing requires four to six workers, and planting/seeding young rice plants requires seven to 12 workers. Threshing can be done by a machine, but it still requires workers to collect it, to bag it, and to perform similar tasks, which takes about one and a half hours per *feddan*. Seeding is manual

MAP 33 *Musharaka by farmers No. 2, No. 6, No. 7, No. 10, No. 17, and No. 28 in Well No. 3 Irrigation District* (2008). *Source*: See Figure 14–5.

work that takes place two months after planting in July or August. The following is an example for rice (Table 63).

Amin Ali Said called 15 men for one day's work of rice seeding/planting in July 2007 from 6 am until 2 pm. These 15 men included relatives such as the son of his sister, and his sister's husband, four men who were waged workers, farmers in the same subdistrict, and some friends and neighbours. Four relatives and friends worked in *musharaka mujamula* (meaning friendly co-operation) without recompense. The four waged workers received 50 Egyptian pounds per day. The other seven people participated in return for help in their fields.

Like other cases of *musharaka* described in Chapter 7, the case of A reveals that the villagers organize co-operation by mobilizing diverse networks, of which family members are the most important, but also friends and neighbours. It is also understood that reciprocity and a wage relationship are not contradictory but can be combined.

In conclusion, the farmers of Well No. 3 Irrigation District manage the lands by combining kinds of *musharaka*. *Musharaka* is a custom that includes many

TABLE 63 *Example of musharaka in rice planting/seedling (July 2007)*

	Age	Primary profession	Relation	Location of residence	Way of *musharaka*
Cultivator	32	Farmer		Izba Othman	
1	34	Farmer	From same family	Izba Othman	*Musharaka mujamla* (friendly cooperation)
2	39	Farmer	From same family	Izba Othman	*Musharaka* mujamla (friendly co-operation)
3	69	Farmer	Neighbour of residence/friend	Izba Othman	*Musharaka* by rotation
4	59	Farmer	Neighbour of residence/friend	Izba Othman	*Musharaka* by rotation
5	49	Farmer	Neighbour of residence/friend	Izba Othman	*Musharaka* by rotation
6	49	Farmer	Neighbour of the Well No. 3 district	Rashda	*Musharaka* by rotation
7	45	Farmer	Neighbour of residence/friend	Izba Othman	*Musharaka* by rotation
8	39	Farmer	Neighbour of residence/friend	Izba Othman	*Musharaka* by rotation
9	24	Farmer	No relation	Rashda	Wage (50 LE per day)
10	25	Farmer	No relation	Rashda	Wage (50 LE per day)
11	26	Farmer	No relation	Arab (Bedouin living near)	Wage (50 LE per day)
12	42	Farmer	Neighbour of residence/friend	Izba Othman	Wage (50 LE per day)
13	40	Farmer	From same family	Izba Othman	*Musharaka mujamla* (friendly co-operation)
14	37	Farmer	From same family	Izba Othman	*Musharaka mujamla* (friendly co-operation)
15	70	Farmer	Neighbour of residence/friend	Izba Othman	*Musharaka* by rotation

Source: Information collected by informant in summer 2007.

types of co-operation. A jaala contract, in a sense, is a type of *musharaka*. In fact, people involved with a jaala are called *shuraka'* (sing. *sharik*) as well as *muja'alin*, as mentioned in Chapter 8. However, only a few jaala contracts are found in Well No. 3 Irrigation District compared with other districts irrigated by local wells, because of the strong intervention or support of the state in water and cultivation management in the government well districts.

Musharaka includes three kinds of co-operation. The first type is co-operation among family members, as discussed in this chapter. The second type is co-operation based on contracts for share-cropping and leasing (*ijar*), as also discussed in this chapter as well as in detail in Chapters 6, 11 and 12. The third type is co-operation in exchange for labour between the villagers, as discussed in this chapter and in Chapter 8.

The farmers of the Well No. 3 Irrigation District use these three types of *musharaka* properly for their land cultivation and management. The common factor among the three kinds of *musharaka* is the co-operation based on a kind of "project" rather than on an institution. It is short-term co-operation applied "on the spot" to any given circumstances.

Conclusion

Rashda is the biggest village in terms of population after Mut, the administrative centre of the Dakhla Oasis region, but it is a so-called new village with a history that extends only as far back as the 19th century. In the 18th century, the area of present-day Rashda was agricultural land, and it was recognized as fields (*ghayt*), related to Ain al-Rashda (Rashda Spring) and belonging to Qalamun, one of the administrative and judicial centres in the Dakhla Oasis region.

Rashda as a settlement was formed in the 19th century by migrants from outside, particularly from Qalamun, the village to which the field of Rashda belonged. This indicates that Rashda was relatively abundant in water and attracted people from other villages to settle in Rashda. The migration was apparently caused by an increase of population and the necessity for land for cultivation.

The most remarkable feature in the history of the villages of the oasis region was the expansion of the agricultural land outside the residential area of the villages. In this sense, Rashda is a village that was born in the modern historical context of the oasis region, and its history positively or negatively reflects the features peculiar to oasis villages in modern times.

Three Factors for Analysis

We began the discussion by defining the society of Rashda as being characterized by three factors: the uncertainty of its water supply, its dependence on the political regime and the rational behaviour of individuals. The purpose of our research on Rashda is to reconstruct the village of Rashda in totality, in historical and contemporary terms. For this purpose, we collected many kinds of source materials on various subjects. This book was mainly written using data and information on the above three factors.

Uncertainty of Water Supply

By "uncertainty", we mean the circumstances that condition and restrict the social and economic life of Rashda. In practice in the context of the oasis region, water is an important factor. Oasis villages, including Rashda, are essentially based on water.

Water is an essential factor for villages in the Nile Valley as well. However, they depend on the Nile, so the availability of water can be controlled and

 | DOI 10.1163/9789004317390_017

estimated every year, unless there is an unexpected catastrophe. In contrast, in the oasis region, where life is totally dependent on underground water, it is difficult to measure the volume of water and to manage the water supply constantly in the medium to long term, because such resources are uncontrollable.

Hydrogeological data on underground water show its relative abundance in Rashda, because the ground-water moves in all directions towards Rashda village. It is also evident in the custom of cultivating rice, a crop that needs abundant water for cultivation.

However, it should be noted that the water capacity is limited, because the ground-water in Rashda is mainly fossil water. There is a possibility that a sudden decrease or lack of water will occur. In fact, all the springs and wells dug before 1960 have now dried up. The uncertainty of water supply is a result of the relationship between supply and demand, and specifically of water, population, and land use. The extension of cultivated land is connected with the availability of water and the location of wells and springs. The farmers' interest lies in the quantity of water rather than in the size of their plots.

Dependence on the Political Regime

Given the level of available technology, the level of underground water will become ever lower as long as it is fossil water. Moreover, the state is required to sustain the water supply for cultivation, because it is very costly to dig a deep well to access underground water.

It is certain that many "investment" wells have been dug by private capital since 1995. However, this so-called horizontal expansion from digging shallow wells will reach a limit when it is no longer possible to obtain water from shallow wells. To avoid such limitations, it is necessary to search for deep water and to dig deep wells—so-called vertical expansion. Huge investments of capital and advanced technology are needed for this.

In developing countries such as Egypt, where private enterprises are not developed, nobody has responsibility for digging deep wells except the state, which has huge amounts of capital and can use advanced technology. Therefore, it is inevitable that the state is involved in water affairs in the oasis region. Thus, although the region may appear to be on the periphery, its survival largely depends on the general political and economic regime of the day.

The involvement of the state in water affairs in Rashda began at the beginning of the 20th century, when availability of useable water was decreasing remarkably. It was promoted by politicians from the 1960s, when the New Valley Project was planned and executed.

The Rational Behaviour of Individuals

However, the state is not an all-round player on this point. The infrastructure for water affairs, which is implemented by the state, can be efficiently used only when the villagers manage it on a proper and rational basis.

In fact, the villagers have developed local customs in water management and crop cultivation that are adapted to the natural environment of Rashda. The villagers have been cultivating crops in ways that they have developed through experience. This local custom is called jaala. Jaala means a legal contract to pay for labour provided in return.

Water affairs, from the digging and management of a well to water distribution, was administered through jaala contracts in the past. People concerned with the contract are called *mutaja'ilin* or *shuraka', which* means the people in charge of, or partners in, the jaala contract. In a word, the jaala is a kind of co-operative project whereby villagers offer labour to each other under circumstances of restricted capital and labour. People who offer labour receive remuneration in return for their work. All the conditions of the project was written in the jaala document. This custom is still commonly used today.

Regardless of these arrangements, it seems that given the level of local technology in Rashda, the management of water affairs is inefficient by the standards of developed countries. For example, the way that water is drained has an important effect on irrigation agriculture, and it is necessary to prevent the soil from becoming too saline. Yet soil salinization is a phenomenon that appears here and there in the irrigated areas of the village.

However, that does not mean that the farmers are not aware of the problem of salinization. Rather, the farmers consider the problem of soil salinization in relation to the costs needed to resolve the issue. They consider the quality of soil, the availability of capital, and the quantity of water, so as to select the most efficient method of irrigation. A jaala is an efficient method of water and land management, being accompanied by other kinds of co-operation in the provision of labour called *musharaka*.

The Case of Well No. 3 Irrigation District

The Well No. 3 Irrigation District, discussed in detail in Part 4, is a suitable place to examine the three factors mentioned above: the uncertainty of water supply, the dependence on the political regime, and the rational behaviour of individuals.

The Well No. 3 Irrigation District is irrigated by government wells exploited by the state, because the district is located in an infertile area of Rashda. This means that the cost of digging deep wells was borne by the state, and the lands belonging to the district were divided into three subdistricts and distributed among cultivators on the principle of equality.

However, once a well is dug and land is distributed by the state, water distribution and the cultivation of the land irrigated by the well are managed by cultivators, although the state is ultimately responsible for the volume of water distribution. Cultivators manage the irrigation and cultivation to adapt to the natural conditions of water and land.

In other words, they consider the quantity of water, the quality of soil, and the availability of capital and labour to select the most efficient method of cultivation and to promote the local customs of cultivation. This explains the productivity and the adoption of crop rotation by the farmers in the district, as demonstrated by our analysis of crop rotation (Kato et al. 2010: 31–34).

It is supposed that the productivity is caused by a complex interaction of many factors, such as the altitude of the land, the distance from a well, soil salinization, the water level, and the quality of social relations in the village and within families.

A Society Embedded in Water Management

We conclude the preceding discussion by modelling the society of Rashda as one based on water, referring to the theory of Karl Polanyi of substantive economy.

Polanyi's Theory of Substantive Economy

As is well known, in the opinion of Karl Polanyi (Polanyi 1977), the word "economic" is used in two different senses: the formal sense and the substantive sense. In the formal sense, "economic" means to "economize" or to be "economical"; that is, to choose among alternatives for the purpose of maximizing output, profit, gaining in exchange, or minimizing the cost of producing something within the context of material "scarcity" relative to what economists call "demand".

In the substantive sense, on the other hand, "economic" simply means the arrangements for acquiring, producing, or using material goods or services for individual or community purposes. According to this meaning, all societies, whatever their size, technology, or political structure, have an "economic" system; that is, structured arrangements for the provision of livelihoods.

In a capitalist, market-integrated economy and in conventional economic theory, these two meanings of "economic" are fused, because in capitalism, market institutions serve both to provide the material means of existence and to enforce economizing activities by the participants. In other words, people must follow the rules of the market to earn their livelihood.

Economic theory reflects this separation of the economy from other social institutions by making market transactions almost its sole concern. However, the market economy is a very special case, historically and anthropologically. Pre-industrial societies frequently have economies in which the structured mode of providing the means of existence does not consist of economizing institutions.

For this reason, Polanyi criticized conventional economic theory, because it perceives all economies, including the primitive and the archaic, as variants of modern industrial economies and translates all economic institutions into market-economizing terms.

Categories of the Economic Integration: Exchange, Reciprocity and Redistribution

Instead of conventional economic theory, Polanyi sought new conceptual categories that would permit both the analysis of the relationship of economic with social organizations and the direct comparison of economies. Economic integration is the key concept of analysis proposed by Polanyi for this purpose, which is applicable to both the market economy and the non-market-integrated economy; that is, the economy is embedded in society in Polanyi's terms.

In his theory, forms of economic integration determine the institutionalized movements through which the elements of the economic process—from material resources and labour to the transportation, storage, and distribution of goods—are connected. Besides exchange, which is the dominant form of integration in our capitalistic society, the other main forms found in human economy are reciprocity and redistribution.

Reciprocity is the obligatory mutual gift-giving between persons who stand in some socially—and especially symmetrically—defined relationship to one another. The key point is that the social relationships involved are paramount and definitive, and these are face-to-face, familiar, personal relationships. The closest term in our current social science terminology is "community".

Redistribution is a movement towards a centre and out of it again, whether the objects are physically moved or only the usufruct over them is shifted. The key point is that centricity is a necessary institutional pattern for the operation of redistribution, and the movement cannot proceed without an established centre from which the redistribution takes place.

Reciprocity is the dominant form of integration in communal societies such as tribes. Redistribution is the dominant form of integration in politically centralized primitive and archaic societies. However, it must be pointed out that the forms of integration proposed by Polanyi are not evolutionary stages, although some did develop earlier than others. Neither are the forms mutually exclusive. In any economy, two, or sometimes three, forms of integration are usually present, although one form tends to be dominant.

A Society Based on Co-operation

In terms of Polanyi's theory, the history of Rashda in modern times could be roughly divided into three periods from the perspective of water affairs. The turning points are the beginning of the 20th century, the 1960s, and the 1990s. The first period until the beginning of the 20th century is the period when water affairs were mainly managed based on a kind of reciprocity on the level of local technology, although differences in property, water, and land among people were observed. The jaala custom can be represented as a kind of reciprocal system to integrate villagers into a joint project.

The second period was the 20th century until the 1960s. At the beginning of the 20th century, the intervention of the state into water affairs began. The measurement of springs by the authority was first conducted in 1902, and permission for digging wells began to be issued in 1910, in the face of a remarkable decrease in water supply. This means that the water infrastructure could not be implemented and maintained without the support of the state in those days. The New Valley Project should be considered to be the turning point of the direct intervention of the state in water affairs. The project redistributed property—water and land—by force.

The third period was from 1995 until the present. The year 1995 was when the law to promote the development of the desert by private capital was passed. As a result, many "investment wells" have been dug outside the cultivated lands of the village for commercial crops such as animal fodders and vegetables. Differences in assets and income would have become wider as a result of the expansion of the exchange economy. However, as mentioned before, the horizontal expansion brought about by digging shallow wells was limited, and capital and technology were required to dig a deep well. The support of the state is indispensible for vertical expansion of agriculture.

The above three periods in the modern history of Rashda, which are classified according to their significance to water affairs, almost exactly correspond to the three periods in the history of village administration: the age of the umda for the first period, the age of the Local Unit for the second period, and the age of privatization for the third period. From the correspondence, we understand the linkage between three factors: the uncertainty of the water supply, dependence on the political regime, and the rational behaviour of individuals. In a sense, these are three categories of economic integration: reciprocity, redistribution, and exchange in terms of Polanyi's theory of the substantive economy.

Select Bibliography

Primary Sources

As mentioned in Introduction, this book uses various kinds of data and information: 1. Agglomerated statistical data, 2. Micro-level data on household, 3. Geographical information, 4. Local documents, and 5. Other qualitative information. Of these, 2, 3, 4, and 5 were collected by field surveys in Rashda.

The main primary sources, on which this book is based, are 2, 3 and 4. As for 2 and 3, they are individually referred to in the text. As for 4, We have collected a lot of local documents possessed by village notables during field surveys in Rashda. Below are the documents that are translated into English or explained in detail on their contents. The first number is Chapter's

List of Documents

5–1 Document dated 12 Rajab 1206 (1792) on the debt of a farmer.

5–2 Document dated 10 Rabiʿ II 1215 (1800) on land lease.

5–3 Document dated Shaʿban 1226 (1811) on the arbitration between two parties on the water distribution of a spring.

5–4 Document dated 23 Shaʿban 1268 (1852) of a *waqf* document, which sheds light on the situation of the fields of Rashda in the first half of the 19th century.

5–5 Document dated 28 Jumada II 1279 (1862) on the arson in "old" Rashda.

6–1 Document dated Safar 1270 (1853/54) on the circumstances in which the villagers of Rashda lived.

7–1 Document dated 24 Dhu al-Hijja 1266 (1850) on the jaala contract for the digging of the new Ain al-Shishlana.

7–2 Document dated 13 July 1937 on the petition of the umda for the digging of a new well because the existing wells in Rashda had dried up.

7–3 Document dated 1 Rabiʿ II 1256 (1840) on the land accumulation by village notables.

7–4 Document dated Dhu al-Qaʿda, 1326 (1908) on the procedure for the digging and completion of a well by jaala contract.

7–5 Document dated Dhu al-Qaʿda, 1326 (1908) on a jaala contract for the repair of a canal called Ain al-Najjarin in Balat.

9–1 Document dated 9 Jumada II 1278 (1861) of the *daftar taʿdad nufus hissat al-shaykh ahmad al-hajj muhammad ʿabd al-hafiz.*

Secondary Sources

Al-Tayb, Sallah. 1985. *Tarikh al-qaba'il al- misriya*. al-Matba'a al-'Asriya, Alexandria.

Antoun, Richard and Iliya Harik (eds) 1972. *Rural Politics and Social Change in the Middle East*. Indiana University Press, Bloomington/London.

Azadiyan. 1931. *Taqarir wa mudhakirat ma'amil maslahat al- sihha al-'umumiya 'an miyah al-sharb fi misr*. Vol. 2, al-Matba'a al-Amiriya, Cairo.

Badawy, H.A.. 2005. "Exchange of Irrigation Water Between Farmers in Egypt". Ninth International Water Technology Conference, Sharm El-Sheikh, Egypt <http://iwtc.info/2005_pdf/07–3.pdf>.

Baer, Gabriel. 1969. *Studies in the Social History of Modern Egypt*. Chicago University Press, Chicago and London, 259p.

Baer, Gabriel. 1976. *Population and Society in Arab East*. Connecticut, Greenwood Press.

Barakat, 'Ali. 1991. "Intifadat al-fallahin fi misr al-hadith 1769–1952", (in Japanese).

Beadnell, H.J. Llewellyn. 1909. *An Egyptian Oasis. An Account of the Oasis of Kharga in the Libyan Desert, With Special Reference to its History, Physical Geography, and Water-Supply*. John Murray, London.

Ben Salem, Leila 1994. "La Famille en Tunisie : Questions et Hypothèses". In *Structures Familiales et Rôles Sociaux*. Actes du Colloque de l'Institut Supérieur de l'Education et de la Formation Continue, 13–27, Cérès Editions, Tunis.

Berque, Jacques. 1955. *Structures Sociales d'un Haut Atlas*. Paris, Presses Universitaires de France, Paris.

Berque, Jacques. 1957. *Histoire Sociale d'un Village Egyptien au XXI ème Siècle*. Mouton, Paris.

Binder, Leonard. 1978. *In a Moment of Enthusiasm: Political Power and the Second Stratum in Egypt*. University of Chicago Press.

Bourdieu, Pierre. 1980. *Le Sens Pratique*. Les Edition du Minuit, Paris.

Brookes, I.A. 2000. "Aeolian Erosional Lineations in the Libyan Desert, Dakhla Region, Egypt". *Geomorphology*. No. 39.

Budyco, M.I. 1956. *Heat Balance of the Earth's Surface*. Seizando, Tokyo.

CAPMAS (Central Agency for Public Mobilization and Statistics). Population Census, various years.

Charrad, Mounira. 1990. "State and Gender in the Maghrib". *Middle East Report*. 163(20:2 March-April).

Charrad, Mounira. 2001. *The Origins of Women's Rights: State and Tribe in Tunisia, Algeria, and Morocco*. University of California Press, Berkeley.

Chaulet, Claudine. 1987. *La Terre, les Frères et l'Argent*. 3 Vols., Office des Publications Universitaires, Alger.

Cuno, Kenneth M. 1992. *The Pasha's Peasants. Land, Society, and Economy in Lower Egypt, 1740–1858*. Cambridge University Press.

Cuno, Kenneth M. 1999 "A Tale of Two Villages: Family, Property, and Economic Activity in Rural Egypt in 1840s", *Proceedings of the British Academy*, 96.

Cuno, Kenneth M. 2015. *Modernizing Marriage. Family, Ideology, and Law in Nineteenth and Early Twentieth Century Egypt.* Syracuse University Press.

Dabous, A.A. and Osmond, J.K. 2001. "Uranium Isotopic Study of Artesian and Pluvial Contributions to the Nubian Aquifer, Western Desert, Egypt". *Journal of Hydrology*. No. 243, pp. 242–253.

Diab, M.Sh.. 1984. "Regional Hydrogeological Reading on the Artesian Basins in North African with Emphasis on the Nubian Aquifer in South Egypt". Egyptian Geological Survey Proceedings of the 3rd Annual Meeting, Cairo.

Division for Public Administration and Development Management (DPADM), Department of Economic and Social Affairs (DESA), United Nations. 2004. *Arab Republic of Egypt. Public Administration Country Profile.*

Duvignaud, Jean. 1968. *Chebika*. Editions Gallimard, Paris.

Dyer, Graham. 1997. *Class, State and Agricultural Productivity in Egypt: Study of the Inverse Relationship between Farm Size and Land Productivity*. Frank Cass, London.

Edmonstone, Sir Archibald. 1822. *A Journey to Two of the Oases in Upper Egypt*. John Murray, London.

"Egypt to Create 3 New Governorates" 2014 Ahram Online 19 August 2014 <http://english.ahram.org.eg/News>, accessed on 16 May 2015.

Fakhry, Ahmed. 1974. *The Oases of Egypt*. Vol. 2, Bahriyah and Farafra Oases. The American University in Cairo Press, Cairo.

Fargues, Philippe. 2003. "Family and Household in Mid-Nineteenth-Century Cairo". In *Family History in the Middle East. Household, Property, and Gender*. Beshara Doumani (ed.), State University of New York Press, Albany.

Food and Agriculture Organization of the United Nations (FAO) website: Egypt, National and local institutions enforcing land regulations <http://www.fao.org/gender-land rights-database>.

Gad, M.I., El-Sheikh, A.E., El Osta, M.M.. 2011. "Optimal Management for Groundwater of Nubian Sandstone Aquifer in Dakhla Depression, Western Desert, Egypt". *International Journal of Water Resources Environment Engineering*. Vol. 3, No. 14.

Garcier, Romain J. and Jean-Paul Bravard. 2014. "Qu'est-ce qu'une oasis? Réflexions géographiques sur un objet-limite". In *Le Myrte et la Rose – Mélanges offerts à Françoise Dunand*, Tallet et Zivie-Coche (dirs.), Presses Universitaires de Montpellier, Montpellier.

Geertz, Clifford, Hildred Geertz and Lawrence Rosen. 1979. *Meaning and Order in Moroccan Society, Three Essays in Cultural Analysis*. Cambridge University Press, London et al..

Gellner, Ernest and John Waterbury. 1977. *Patrons and Clients in Mediterranean Societies*. Gerald Duckworth and Co, London.

Hansen, Bent. 1975. "Chapter 6: Basic Characteristics of Egyptian Agriculture". In *Foreign Trade Regimes and Economic Development: Egypt*. Hansen, Bent and Karim Nashashibi (eds.), NBER <http://www.nber.org/books/hans75–1>.

Herodotos. *The Histories*. Volume 3 <http://perseus.uchicago.edu/perseus-cgi>, accessed on 16 May 2015.

Himida, I.H. 1966. *A Quantitative study of the Artesian Water Exploitation Resources* In *Kharga and Dakhla Oases, Western Desert of Egypt*. Desert Research Institute Bulletin, Egypt.

Hopkins, Nicholas. 1993. "Small Farmer Households and Agricultural Sustainability". In, *Sustainable Agriculture in Egypt*. Faris, M.A. and M.H. Khan (eds), Lynne Rienner, Boulder and London.

Hopkins, Nicholas S. (ed.) 2003. *The New Arab Family*. Cairo Papers in Social Science, Vol. 24, No. 1/2.

Horden, Peregrine and Nicholas Purcell. 2000. *The Corrupting Sea: A Study of Mediterranean History.* Blackwell Publishers.

Hunayn, J. 1904. *Al-atyan wa al-dara'ib fi al-qutr al-misri*, al-Matba'a al-Kubra al-Amiriya, Bulaq (Cairo).

Husayn, 'Alaya Hasan. 1975. *Al-wahat al-kharijiya. Dirasat fi al-tanmiya wa al-taghayyur al-ijtima'i fi al-mujtama'atal-mustaddathat*, Al-Hy'a al-Misriya al-'Amma li-l-Kitab, Cairo.

Ibrahim, Fouad N. and Barbara Ibrahim. 1993. *Egypt: An Economic Geography*. I.B. Tauris, London/New York.

Ibrahim, Solava. 2006. *The Role of Local Councils in Empowerment and Poverty Reduction in Egypt*. Cairo Papers in Social Science Vol. 27, No. 3, The American University in Cairo Press, Cairo.

"Irrigation Ministry Prevent Rice Cultivation in Deserts". 2004 In *Egypt Independent* 2014/4/24 <http://www.egyptindependent.com/news>.

Iwasaki, Erina. 2006. "Analytical Framework for the Analysis of Kinship in North African Rural Societies: a Case Study of Commercial Migration in Southern Tunisia". *Mediterranean World.* Vol. 18, the Mediterranean Studies Group, Hitotsubashi University.

Iwasaki, Erina. 2007. "What is the Aila? : A Comparative Study of Kinship Structure in Egyptian Villages." *Annals of Japan Association for Middle East* (*AJAMES*). Vol. 22, No. 1.

Kato, Hiroshi. 1980. "Egyptian Village Community under Muhammad 'Ali's Rule – An Annotation of Qanun al-Filaha". *Orient*, Vol. XVI.

Kato, Hiroshi. 1990. "Nomads and Farmers in the Process of the Modernization of Egypt." *Orient*, Vol. XXVI.

Kato, Hiroshi. 1993. *Private Landownership and Egyptian Society.* Soubun-sha, Tokyo (in Japanese).

Kato, Hiroshi. 1995. "The Egyptian Rural Society in the Mid-Nineteenth Century as Reflected in the Document on Petition for the Exemption from Military Service". *Mediterranean World.* Vol. 14, The Mediterranean Studies Research Group ed., Hitotsubashi University.

Kato, Hiroshi. 2001. "The Bedouin in Egyptian National Identity: Minority or Vagabond?" In *State Formation and Ethnic Relations in the Middle East.* Usuki Akira (ed.), JCAS Symposium Series 5, the Japan Center for Area Studies (JCAS), National Museum of Ethnology, Osaka, Japan.

Kato, Hiroshi. 2008. "Is the Egyptian Village a Community." *International Journal of Public Affaires*, Chiba University, Vol. 4, Research Center on Public Affairs for Sustainable Welfare Society, Chiba University.

Kato, Hiroshi and Erina Iwasaki. 2008. "Rashda. A Village in Dakhla Oasis". *Mediterranean World.* Vol. 19.

Kato, Hiroshi and Erina Iwasaki. 2011. "Egyptian Village Map". *Hitotsubashi Economics.* Vol. 4, No. 1, Hitotsubashi University (in Japanese).

Kato, Hiroshi, Iwasaki, Erina, Nagasawa, Eiji, Anyoji, Hisao, Matsuoka, Nobuhiro and Kimura, Reiji. 2010. "Rashda: System of Irrigation and Cultivation in a Village in Dakhla Oasis". *Mediterranean World.* Vol. 20.

Kato, Hiroshi, Reiji Kimura, Salwa F. Elbeih, Erina Iwasaki, El-Sayed A. Zaghloul. 2012. "Land Use Change and Crop Rotation in Rashda Village, Dakhla Oasis, Egypt: Analysis of a Government Well District Based on Satellite Data". *Mediterranean World.* Vol. 21.

Kato, Hiroshi, Reiji Kimura, Salwa F. Elbeih, Erina Iwasaki, and El-Sayed A. Zaghloul. 2012. "Land Use Change and Crop Rotation Analysis of a Government Well District in Rashda Village– Dakhla Oasis, Egypt Based on Satellite Data". *Egyptian Journal of Remote Sensing and Space Sciences.* Vol. 15, No. 2.

Kato, Hiroshi, Hiroomi Tsumura, and Erina Iwasaki. 2013. "GIS as a Tool for Researching the Socioeconomic History of Modern Egypt". *Journal of Asian Network for GIS-based Studies.* No. 1.

Kato, Hiroshi, Salwa Elbeih, Erina Iwasaki, Ahmed Sefelnasr, Adel Shalaby, and Elsayed Zaghloul. 2014. "The Relationship between Groundwater, Landuse, and Demography in Dakhla Oasis, Egypt". *Journal of Asian Network for GIS-based Historical Studies.* No. 2.

Kenneth, P. Beveridge and W.W. Mackie. 1925. *Berseem or Egyptian Clover (Trifolium Alexandrinum) : A Preliminary Report.* Berkeley, Cal.: Agricultural Experiment Station <http://catalog.hathitrust.org/Record/100089496>.

Kilani, Taha Muhammad. 1994. *Qabila al-jawazi fi mawkib al- tarikh.* Dar Hara', Minya.

Kimura, Yoshihiro. 1973. "The Structure of Egyptian Rural Society before the Agrictulrual Reform". In *Historical Studies of Community.* Kawashima, Takeyoshi and Kazuhiko Sumiya (eds), Institute of Developing Economies, Tokyo (in Japanese).

King, W.J. Harding. 1917. "Irrigation in the Dakhla Oasis". *The Geographical Journal.* Vol. 50, No. 5.

Lozach, J. and Hug, G. 1930. *L'habitat rural en Egypte*. L'Institut Français d'Archéologie Orientale du Caire, Cairo.

Lyons, H.G. 1908. *The Cadastral Survey of Egypt. 1892–1907*. Ministry of Finance, National Printing Department, Cairo.

Mayfield, James B. 2012. *Field of Reeds. Social, Economic and Political Change in Rural Egypt: In Search of Civil Society and Good Governance*. AuthorHouse, Bloomington.

Mehenna et al. 1994. *Farmers and Merchants: Background to Structural Adjustment in Egypt*. Cairo Papers in Social Science. Vol. 17, No. 2 (summer).

Ministry of Health and Population, El-Zanaty and Associates, ICF International (EDHS). 2014. *Egypt Demographic and Health Survey 2014. Main Findings* <http://dhsprogram.com/pubs/pdf/PR54/PR54.pdf>.

Ministry of Water Resources and Irrigation. Ministry website <http://www.mwri.gov.eg/historyofministry.aspx>.

Miyaji, Mieko. 1993. "Introduction to the Anthropology of Migration". *Applied Sociological Studies*. Vol. 3 (February) (in Japanese).

Montaz and Siddiq. 1989. "Extending Rice Cultivation to New Areas in Egypt". In *Rice Farming Systems. New Directions.* International Rice Research Institute, Proceedings of an International Symposium 31 January – 3 February 1987, Rice Research and Training Center, Sakha, Egypt. <http://pdf.usaid.gov/pdf_docs/pnabd499.pdf>, accessed on May 22 2015.

Morris, Ellen F.. 2010. "Insularity and Island Identity in the Oases Bordering Egypt's Great Sand Sea". In *Thebes and Beyond: Studies in Honour of Kent R. Weeks*. Zahi Hawass and Salima Ikram (eds.). Supreme Council of Antiquities Press, Cairo <http://www.academia.edu/373128>.

Nakaoka, Saneki. 1973. "Community in Egypt: Note on the Forms and Agents of the Appropriation of Property". In *Historical Studies of Community.* Kawashima, Takeyoshi and Kazuhiko Sumiya (eds), Institute of Developing Economies, Tokyo (in Japanese).

Nizarat al-Maliya. 1909. *Ihsaiya ʿumumi ʿan muhafazat wa al-mudiriyat al-quttr al-misri*, Cairo.

Peters, Rudolph. 2008. "Sharecropping in the Dakhla Oasis. Shariʿa and Customary Law in Ottoman Egypt" In *The law applied: Contextualizing the Islamic Shariʿa*. P. Bearman, W. Heinrichs and B. Weiss (eds.), A volume in honor of Frank Vogel. I.B. Tauris, London.

Peters, Rudolph. 2011. *Wathaʾq madinat al-qasr bi-l-wahat al-dakhla: masdaran li-taʾrikh misr fi-l-ʿasr al-ʿuthmani,* Silsilat Dirasat Wathaʾiqiya, Matbaʿat al-ʿAmma li-Dar al-Kutub wa-l- Wathaʾq al-qawmiya, Cairo.

Polanyi, Karl. 1977. *The Livelihood of Man*. Edited by Harry W. Pearson. Academic Press.

Sampsell, Bonnie M. 2003. *The Geology of Egypt. A Traveller's Handbook*. The American University in Cairo Press, Cairo.

Ramzi, Muhammad. 1968. *al-Qamus al-jughrafi lil-bilad al-misriya: min 'ahd qudama' al-misriyin ila sanat 1945*, 4vols, Dar al- Kutub, Cairo

Rennel, James. 1830. *The Geographical System of Herodotus Examined and Explained, by a Comparison with Those of Other Ancient Authors, and with Modern Geography*. Vol. 2, C.J.G. & F. Rivington, London.

Richards, Alan. 1982. *Egypt's Agricultural Development, 1800–1980. Technical and Social Change*, Westview Press.

Robaa, S.M. 2008. "Evaluation of Sunshine Duration from Cloud Data in Egypt". *Energy*. No. 33.

Saad, Reem. 1988. *Social History of an Agrarian Reform Community in Egypt*. Cairo Papers in Social Science, Vol. 11, No. 4.

Sefelnasr, Ahmed. 2008. Development of Groundwater Flow Model for Water Resources Management in the Development Areas of the Western Desert, Egypt. PhD Thesis, Martin Luther University Halle-Wittenberg, ULB Sachsen-Anhalt <http://sundoc.bibliothek.uni-halle.de/diss-online/07/07H178/prom.pdf>.

Singerman, D. and H. Hoodfar. 1996. *Development, Change, and Gender in Cairo: a View from the Household*. Indiana University Press, Bloomington & Indianapolis.

Singerman, Diane. 1997. *Avenues of Participation: Family, Politics, and Networks in Urban Quarters of Cairo*. The American University in Cairo Press, Cairo.

State Information Service "Local Administration" <http://www.sis.gov.eg>, accessed on 16 May 2015.

Sykes, Godfrey. 1912. "A Journey in the Libyan Desert". *Bulletin of the American Geographical Society*. Vol. XLIV, No. 10.

Tadros, M.T.Y.. 2000. "Uses of Sunshine Duration to Estimate the Global Solar Radiation over Eight Meteorological Stations in Egypt". *Renewable Energy*. No. 21, pp. 231–246.

United Nations Environment Program (UNEP). 1997. *World Atlas of Desertification*. Arnold, London.

Viscomi, Joseph. 2010. *The Farthest Place. Social Boundaries in an Egyptian Desert Community*. Cairo Papers in Social Science, Vol. 30, No. 2, The American University in Cairo Press, Cairo.

Vivian, Cassandra. 2008. *The Western Desert of Egypt*. New Revised Edition, The American University in Cairo Press, Cairo.

Voll, Sarah P. 1980. "The Egyptian Land Reclamation Since The Revolution". *The Middle East Journal*. Vol. 34, No. 2.

Walz, Terence. 1978. *Trade between Egypt and Bilad as-Sudan 1700–1820*. Institut Francaise d'Archedologie Orientale du Caire, Cairo.

Wilkinson, Gardner. 1847. *Murray's Hand-Book for Travellers in Egypt*. Spottiswoode and Show, London.

Willams, Captain Claud H. 2013. *Light Car Patrols 1916–19. War and Exploration in Egypt and Libya with Model T Ford.* Silphium Press, London.

Zanaty, Fatima and Ann Way. 2009. *Egypt Demographic and Health Survey 2008.* Ministry of Health, El-Zanaty and Associates, and Macro International, Cairo.

Index of Names

General Index